Statistics in
Market Research

Arnold Applications of Statistics Series

Series Editor: **BRIAN EVERITT**
Department of Biostatistics and Computing, Institute of Psychiatry, London, UK

This series offers titles which cover the statistical methodology most relevant to particular subject matters. Readers will be assumed to have a basic grasp of the topics covered in most general introductory statistics courses and texts, thus enabling the authors of the books in the series to concentrate on those techniques of most importance in the discipline under discussion. Although not introductory, most publications in the series are applied rather than highly technical, and all contain many detailed examples.

Other titles in the series:

Statistics in Education   Ian Plewis
Statistics in Civil Engineering   Andrew Metcalfe
Statistics in Human Genetics   Pak Sham
Statistics in Finance   Edited by David Hand and Saul Jacka
Statistics in Sport   Edited by Jay Bennett
Statistics in Society   Edited by Daniel Dorling and Stephen Simpson
Statistics in Psychiatry   Graham Dunn
Statistics in Archaeology   Michael Baxter

Statistics in Market Research

Chuck Chakrapani

Chief Executive Officer of Millward Brown Goldfarb

Adjunct Professor, Michael G. DeGroote School of Business,

McMaster University

A member of the Hodder Headline Group
LONDON
Distributed in the United States of America by
Oxford University Press Inc., New York

First published in Great Britain in 2004 by
Arnold, a member of the Hodder Headline Group,
338 Euston Road, London NW1 3BH

http://www.hoddereducation.com

Distributed in the United States of America by
Oxford University Press Inc.
198 Madison Avenue, New York, NY 10016

The advice and information in this book are believed to be true and
accurate at the date of going to press, but neither the authors nor the publisher
can accept any legal responsibility or liability for any errors or omissions.

British Library Cataloguing in Publication Data
A catalogue record for this book is available from the British Library

Library of Congress Cataloging-in-Publication Data
A catalog record for this book is available from the Library of Congress

ISBN-10: 0 340 76397 3
ISBN-13: 978 0 340 76397 1

2 3 4 5 6 7 8 9 10

Typeset in 10 on 11pt Times by Dorchester Typesetting Group Ltd
Printed and bound in India

What do you think about this book? Or any other Hodder Arnold title?
Please send your comments to www.hoddereducation.com

For
Brian Everitt
Without whose encouragement I would never have finished this book

Contents

Illustrative Marketing Problems

Preface

The purpose of this brief volume is to illustrate how a variety of marketing problems can be solved using standard statistical techniques. Two groups of audience readily come to mind: (1) marketers and marketing researchers who would like to know how a wide variety of marketing problems may be solved using statistical techniques; and (2) statisticians who are not familiar with the marketing applications of statistics. Although the technical aspects of each technique are outlined briefly, the emphasis is on marketing applications rather than on statistics.

Because of the diversity of audience I have in mind for this book, for each technique I have provided a near-non-technical introduction followed by an explanation of how the technique works (including an explanation of the computer output). To illustrate how these techniques can be used in a marketing context, I have included three detailed applications for each core technique. Each application is chosen such that it deals with a different marketing problem.

Let me give a brief explanation of what this book is, what it is not, and how it is different from other books on the market.

This book concentrates on material that helps to bridge the gap between statistics and marketing. This book does not aim to teach statistics to marketers or marketing to statisticians. Rather, it provides a description of statistical techniques making them accessible to any interested professional, even if that person is not statistically inclined. It describes a variety of real-life marketing problems that are difficult to solve without the aid of statistics and shows how the application of statistics helps the marketer to solve the problem effectively. Thus it avoids replicating standard texts on statistics and marketing and concentrates on material that helps to bridge statistics and marketing.

The material presented is unapologetically applied in its orientation. This book concentrates on the applications of statistics that are relevant to a marketer faced with a real-life problem. The material presented here primarily does not concern itself with theory. For instance, marketers who want to group attributes can use principal components analysis. I have not gone into the theoretical issues surrounding this: whether they should group attributes at all, or whether it would be preferable for them to use common factor analysis instead. This is not because these issues are not important, but because they need a depth of understanding that is best left to other sources.

This book illustrates both common and uncommon applications of statistical methods to marketing problems. Books on applied statistics generally illustrate the techniques using contrived, trivial or trite examples. By letting my practical experience in the field help me to avoid the contrived, the trivial and the trite, I have searched for applications that are both common and novel. I believe that this approach will illustrate the power of statistical techniques much more convincingly in solving a variety of marketing problems.

The examples used in this book relate to problems that are frequently encountered in marketing and research. In illustrating how statistics can be used to solve marketing problems, I have avoided using cases that are artificial and situations that are not realistically encountered in the marketing context. I have tried to avoid giving the impression that statistical methods are solutions looking for problems. Even when an application is novel, I have tried to make sure that the marketing problem is real and the technique used is relevant.

The applications included in this book come from all over the world. The illustrative cases in this book are drawn from research carried out in different parts of the world, including Australia, Canada, Germany, the Netherlands, New Zealand, Spain, Sri Lanka, the UK and the USA.

The book should appeal both to applied statisticians and to marketing practitioners. I have provided a brief technical description of the techniques in the Appendix for those who would like to know the statistical underpinnings, and an explanation of computer outputs for those who are interested in solving the problem using packaged computer programs.

Inevitably, different chapters in the book and different sections within a chapter are interdependent. I have tried to keep such interdependence to a minimum, so even if you skip the technical sections and read only the introduction and the marketing cases, you will still be able to follow the materials presented.

Although I have illustrated how marketing problems can be solved through statistical methods, this should not be construed as a cookbook. The examples here are not meant to be templates for solving similar marketing problems since there may be better methods to solve the problem depending on the context. I hope the examples serve as reminders of the vibrancy of statistical methods rather than as sterile templates.

This book is designed to be self-contained and can be used independently of other sources. It can also be used as a useful supplementary text for any of the following subjects: marketing, marketing research, and applied multivariate analysis.

Acknowledgements

I would like to thank Brian Everitt of the University of London for asking me to write this book, Peter Charlton of the University of Portsmouth, John Smart of S.M. Research, Dr T. K. Gopalan, and Nancy Kramarich of Pfizer Consumer Healthcare for reviewing and commenting on the material, Ken Deal for locating some of the case studies used in this book, Richard Leigh for his superb copyediting and Christine Mole for assisting me in all aspects of preparing this book.

For support in developing the book, I would like to thank Liz Gooster, Christina Wipf Perry and Tiara Misquitta, all of Arnold.

For permission to reproduce copyrighted materials I thank the following: the Professional Marketing Research Society for permitting me to reproduce materials from the *Canadian Journal of Marketing Research*, the American Marketing Association for permitting me to reproduce materials from the AMA journals, Professor Philip Gendall for permission to reproduce the modified Fearon charts, and Professor J.D. Hunt for providing me with the actual conjoint utilities on which his paper was based. I would also like to thank the authors and publishers of the many case studies summarized in this book.

About the author

Dr. Chuck Chakrapani is Chief Executive Officer of Millward Brown Goldfarb, Adjunct Professor of Marketing at DeGroote Graduate School of Business at McMaster University and Chairman of the Investors Association of Canada. He has held academic appointments at the London Business School and at the University of Liverpool. He is the editor-in-chief of *Canadian Journal of Marketing Research* and of *Marketing Research*. He has authored more than 10 books including *Marketing Research: State-of-the-Art Perspectives* (ed.) and *Modern Marketing Research Step-by-Step*. Chakrapani is a fellow of the Royal Statistical Society. He was also elected as a fellow of the Professional Marketing Research Society for his 'outstanding contributions to marketing research in Canada'. He can be reached at chuck.chakrapani@ca.mbgoldfarb.com or at chuck@chakrapani.name.

Part 1

Introduction

1

Statistics and Marketing

1.1 Introduction

Statistical methods are used extensively in marketing. Marketing researchers and modellers rely heavily on advanced statistical techniques to solve marketing problems. No area of marketing – be it product or service introduction, promotion, positioning, branding, advertising, segmenting, or sales forecasting – is free of the influence of statistical methods. Yet not every marketing professional is aware of the variety of problems to which statistical methods can be applied. Similarly, many statisticians are not familiar with the way in which their techniques are being used to solve marketing problems. The purpose of this book is to provide a glimpse of the many ways statistics can be used to solve marketing problems.

Statistical methods used in marketing range from simple exploratory data analysis and significance testing to highly complicated mathematical modelling techniques. The methods covered here are mostly multivariate statistical techniques. These techniques are uniquely suited to solving marketing problems. There are many reasons for this.

First, statistical methods help the marketer to summarize efficiently large amounts of data. Marketing research data tend to be voluminous. Multivariate analysis can be used to summarize the data. For instance, consumers may be asked to evaluate a product on 30 dimensions. Techniques such as factor analysis help the marketer to summarize the data in fewer dimensions. Again, a marketer may want to group the consumers as belonging to different segments. Techniques such as cluster analysis provide statistical means of achieving these ends. Overall, the interdependent methods of multivariate analysis help the marketer with data reduction and grouping problems.

Secondly, statistical methods help the marketer to understand the effects of a number of variables on a marketing outcome. How do product quality, service quality, and price affect customer satisfaction? How do economic conditions, price, and inflation affect sales? The dependent methods of multivariate analysis help the marketer solve problems such as these.

Thirdly, statistical methods help the marketer to minimize the confounding effects inherent in most marketing data. In practice, most marketing research data are observational in nature, obtained from databases or from surveys. Even when marketing experiments are set up – for example, using two different types of promotions in two different markets and comparing their effects on sales – the results are still subject to other confounding effects. If the sales in the two markets are different, is the difference due to differences in promotion or differences in some other characteristics of the markets? One can, in theory, select several markets using random sampling procedures to minimize such confounding effects. Unfortunately, this is often too expensive to be practical. So, in reality, when marketing experiments are carried out, they tend not to be as reliable as those in other areas of study such as agriculture or biology.

Because the effects of variables that influence a marketing phenomenon are not taken into account through randomization, it becomes important to study the effect of variables that may potentially affect the results. While statistical analysis is no substitute for well-conducted randomized experiments, taking into account the effect of several variables through statistical methods is better than doing nothing about confounding variables. Obviously, it is not possible to control the confounding effects of unknown variables. Nevertheless, many confounding variables are known in marketing and therefore their effects can be estimated using statistical methods. Multivariate methods are well suited to studying the simultaneous effects of many variables.

Finally, statistical methods enable the marketer to assess the effects of alternative future scenarios. Marketers are expected to make decisions under conditions of uncertainty. What will happen to sales if we drop our price by 10%? How will this be affected if our competitors also drop their price by 10%? By 15%? Statistical methods provide means to model alternative scenarios to enable marketers to choose a course of action among the many that are available.

Although statistical methods are widely used in marketing (especially in marketing research), the level of sophistication varies.

- Basic statistics – the use of means, margins of error, statistical significance, etc. – is nearly universal in marketing, especially in marketing research.
- Multivariate analysis and basic modelling techniques are very widely – but not universally – used in marketing. Techniques such as cluster analysis, perceptual mapping and conjoint analysis have found many applications in marketing.
- Modelling techniques that attempt to apply statistical and mathematical techniques to marketing mix decisions are frequently used in marketing, but less so than multivariate analysis.

This book assumes that the reader already has some familiarity with basic statistics and covers multivariate analysis and simpler modelling techniques, but not advanced modelling techniques. As mentioned in the Preface, the material is structured such that even those who are not familiar with basic statistics can still follow a substantial part of the book, especially the cases.

1.2 Choosing a multivariate technique

The choice of a specific multivariate technique depends on two considerations: What type of data are we dealing with? Are we interested in a grouping of objects (data reduction and relationship clarification) or how certain variables influence other variables ('cause–effect' analysis).

Type of data

Marketing data can be either metric or non-metric. Metric data are measured data, such as income, age, years of schooling. When the data are metric we think in terms of averages: What is the average income? Average age? Average years of schooling? Metric data are also considered continuous and their scale points are equally scaled. For instance, the difference between an income of £20 000 and £30 000 is equal to the difference between an income of £30 000 and £40 000: £10 000 in both cases.

Non-metric data are counted data. Gender, place of residence, and level of education are all non-metric. When the data are non-metric we think in terms of percentages: What proportion of Liverpool residents are women? What percentage of the UK population is Scottish? What proportion of those who attend college were born outside the country? Non-metric data are not considered continuous.

Ordinal data are a special case of non-metric data in which assignment of numbers to categories has intrinsic meaning. Consider the following assignment of numbers:

1	Captain
2	Major
3	Colonel
4	Field marshal

As the scale value increases, so does the rank. However, we cannot infer that the difference between the ranks of captain and major is the same as that between colonel and field marshal. Ordinal data naturally arise in marketing when customers rank different brands in the order of their preference.

In marketing, the distinction between metric and non-metric data is not always clear-cut. If a consumer rates a product on a five-point scale in which 1 stands for least satisfied and 5 for most satisfied, is the scale metric or non-metric? Although the difference between 3 and 4 and between 4 and 5 is the same, one scale point, it may well be that the difference between what 3 represents (neither satisfied nor dissatisfied, say) and what 4 represents (fairly satisfied, say) cannot be compared with the difference between what 4 represents and what 5 represents (two levels of satisfaction). The former difference may be more critical than the latter. Note, however, that in marketing, by convention, all rating scales that have five points or more are considered metric.

Metric data can be presented in a non-metric form. Suppose we collect the actual income of consumers but report it as a percentage of consumers who earn over £30 000 and under £30 000. Our data are metric and our reporting of them is non-metric.

Because metric and non-metric data have different properties, the techniques used to analyse these two types of data are different. However, many metric techniques provide for the inclusion of non-metric data. For instance, multiple regression analysis, which is a metric technique, provides for the inclusion of non-metric variables.

Dependence and interdependence

Traditionally multivariate techniques are broadly grouped into two categories: *interdependent* and *dependent*. Interdependent techniques (also known as grouping or structural techniques) concern themselves with grouping similar objects such as attributes or consumers. For instance, a marketer may be interested in knowing whether there are different market segments with different needs that can be met with the introduction of specific products. Dependence techniques (also known as functional techniques) study the effect of a number of variables on one or more variables. For instance, we may be interested in knowing how different demographic attributes (predictor or independent variables) affect a consumer's intent to buy a product (criterion or dependent variable).

From a technical point of view, in dependence techniques the data matrix to be analysed is partitioned into criterion and predictor subsets. In interdependence techniques the data matrix is not so partitioned.

More complex dependence models have multiple criterion variables and multiple predictor variables. They attempt to analyse the effects of a set of variables (such as demographics) on another set of variables (attitudes towards buying expensive items). Their use in marketing is less widespread, presumably because of the complexities involved in the analysis as well as in the communication of results.

EXHIBIT 1.1

Basic multivariate techniques

Dependence techniques

Dependent variable(s)	Independent variables	Technique
Metric	Metric	*Multiple regression*
Non-metric	Metric	*Multiple discriminant analysis*, logistic regression
Metric	Non-metric	Analysis of variance
Non-metric	Non-metric	*Conjoint analysis*, discrete discriminant analysis
Metric*	Metric	Canonical correlation
Metric*	Non-metric	Multiple analysis of variance
Non-metric*	Non-metric	Discrete multiple discriminant analysis

* Multiple dependent variables.

Interdependence techniques

Variables	Objective	Technique
Metric	Group variables	*Principal components and factor analysis*
Non-metric	Relating two sets of data	*Correspondence analysis* Loglinear models
Metric	Relating two sets of data	Biplots
Metric/ non-metric	Group objects	*Cluster analysis*

The more frequently used techniques in marketing, italicized in Exhibit 1.1, form the core of this book. We also discuss other – somewhat less frequently used – techniques that are considered 'advanced', such as *path analysis, structural equation modelling,* and *data mining techniques.*

Basic data manipulation

Multivariate analysis is concerned with the study of associations among sets of measurements. A typical data set in marketing research consists of individuals (e.g., consumers) and their characteristics (e.g., products used by consumers, consumer attitudes, and demographics) relevant to the marketer. In analysing such data, four computing themes recur:

1. summarizing the data (means, variances and the like);
2. computing sums of squares (for the raw, mean-corrected and standardized data);
3. analysing relationships between variables (covariance and correlation);
4. linearly combining data (creating new variables as a linear combination of the original variables).

Conceptually, these concepts are trite and based on commonly used statistical principles. However, in multivariate analysis, it is more common to express them in matrix notation, since this provides an effective way of describing multivariate data-analytic computations (see Appendix).

1.3   An overview of multivariate techniques

This section provides a brief overview of different multivariate techniques that can be used in marketing.

EXHIBIT 1.2
___

Principal components and factor analysis

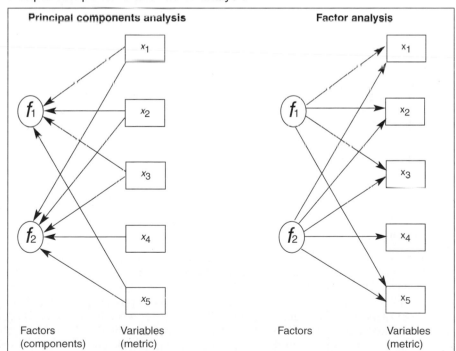

Principal components and factor analysis

Principal components analysis is a metric data reduction technique that reduces a large number of variables into fewer composite variables, with minimum loss of information (by maximizing the amount of variance explained). Each composite so created tends to correlate with or 'load on' variables that are conceptually similar. This provides a logical way of organizing a large number of variables in terms of fewer logical composite variables. For example, customers may rate a company on 25 different variables. Principal components may be able to summarize this data set into just five new variables called factors or components (linear combination of the original variables) such as product quality, service quality, price, ease of contact and reliability that explain, say, 75% of the variance in the original data. If we further assume that the observed variables are an expression of these underlying factors, we have a different model – common factor analysis (Exhibit 1.2).

Correspondence analysis

Correspondence analysis can be viewed as a technique that models contingency tables (known as cross-tabulated data). Often described as principal components of non-metric data, correspondence analysis is frequently used in marketing research to create perceptual maps that relate two sets of data such as demographics and brand usage (Exhibit 1.3).

EXHIBIT 1.3

Correspondence analysis

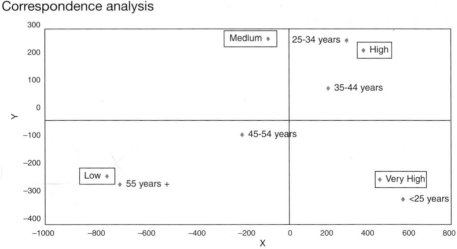

Cluster analysis

Cluster analysis groups individuals into different clusters, based on a given set of variables such as demographics or attitudes. Two basic criteria used to arrive at clusters are: (1) individuals within a cluster should be as close to one another as possible; and (2) individuals of one cluster should be as far away as possible from individuals of other clusters. This technique is helpful to the marketer when he or she intends to identify market segments based on, for instance, consumer attitudes. Factor analysis uses correlations to identify similarities among variables, while cluster analysis uses the similarity of characteristics of the individuals to group them (Exhibit 1.4).

EXHIBIT 1.4

Cluster analysis

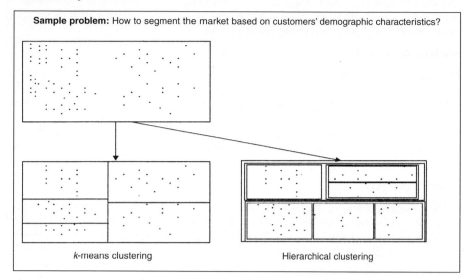

Sample problem: How to segment the market based on customers' demographic characteristics?

k-means clustering   Hierarchical clustering

Multiple regression

Multiple regression is a dependence technique that requires metric data input, although some independent variables can be non-metric. A marketing manager

EXHIBIT 1.5

Multiple regression

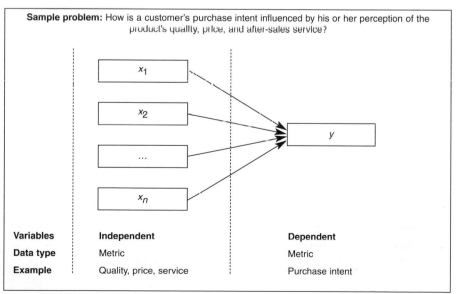

Sample problem: How is a customer's purchase intent influenced by his or her perception of the product's quality, price, and after-sales service?

Variables	Independent	Dependent
Data type	Metric	Metric
Example	Quality, price, service	Purchase intent

interested in knowing which attributes – such as product quality, service quality, after-sales service, or on-time delivery – contribute to purchase intent, and to what extent, can use multiple regression analysis. The input data could be customer evaluations of the company on a 10-point rating scale on the attributes of interest such as product, price, service, and purchase intent (Exhibit 1.5).

Analysis of variance

In some instances the *independent* variables may be categorical. For instance, the marketer may be interested in knowing how income level (high, medium, low), education level (high school, graduate, postgraduate) and gender (male, female) affect overall satisfaction. Since the independent variables are non-metric, we can use analysis of variance rather than multiple regression. Some studies show that using categorical variables in a regression equation produces results that are similar to those obtained using the analysis of variance.

Conjoint analysis

Conjoint analysis and its variants are very popular in marketing research. Consider the analysis of variance problem in which the dependent variables are *ordinal* (ranked as opposed to being truly metric). Suppose a computer manufacturer has identified three attributes, Cost (<1000, 1000–1999, 2000+ dollars), weight (< 4 lb, 4–6 lb, 7 lb+) and built-in drives (yes, no), that customers focus on when buying a computer. What the manufacturer really would like to know is what level of an attribute the

EXHIBIT 1.6

Conjoint analysis

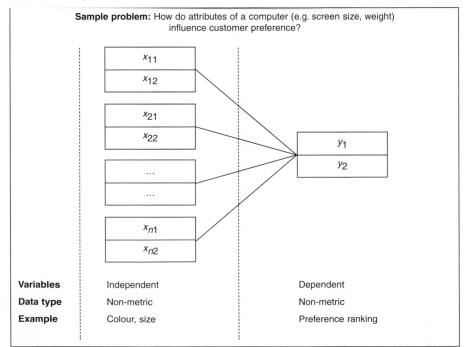

Variables	Independent	Dependent
Data type	Non-metric	Non-metric
Example	Colour, size	Preference ranking

customer would trade for what level of some other attribute. Will the customer pay $2000+ if the computer weighs less than 4 lb? Would the customer accept a slightly heavier computer if it included built-in disk drives?

Typically, different combinations of these features (e.g., cost < $1000, weight 4–6 lb, no built-in disk drives; cost $2000+, weight < 4 lb, built-in disk drives) are presented to consumers who are then asked to rank the different combinations presented to them. The purpose of conjoint analysis is to identify the relative importance of each attribute and attribute levels to the customer (Exhibit 1.6).

Discriminant analysis

Discriminant analysis handles problems that are very similar to those handled by regression analysis except for the fact that the dependent variable is non-metric rather than metric. For instance, if we are interested in knowing which attributes – product quality, service quality, after-sales service, or on-time delivery – contribute to purchase decision (yes or no) and to what extent, we could use discriminant analysis. Multiple discriminant analysis handles problems that involve dependent variables with several categories. In the above example, if we try to predict the importance of different variables in choosing among two different brands, then we could use multiple discriminant analysis (Exhibit 1.7).

EXHIBIT 1.7

Discriminant analysis and logistic regression

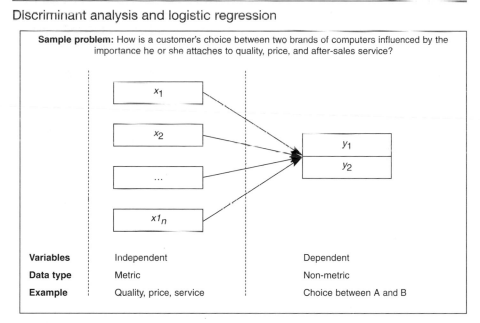

Sample problem: How is a customer's choice between two brands of computers influenced by the importance he or she attaches to quality, price, and after-sales service?

Variables	Independent	Dependent
Data type	Metric	Non-metric
Example	Quality, price, service	Choice between A and B

Logistic regression

Logistic regression can be used as an alternative to discriminant analysis since, in many contexts, these techniques can be used interchangeably to solve the same marketing problems. There are two advantages to using logistic regression over discriminant analysis: (1) logistic regression, unlike discriminant analysis, does not assume that the data came from a multivariate normal distribution; (2) logistic regression also provides, as a standard output, the probability of each individual belonging to a group (e.g., the probability that consumer x is buying our product given her demographics).

Discrete discriminant analysis

When both dependent and independent variables are non-metric, then we use discrete discriminant analysis. An example of this would be trying to predict whether a consumer is likely to buy brand A or brand B, given his income (high, medium, low), educational level (high school, college graduate, postgraduate) and gender (male, female). Discrete discriminant analysis, which uses the rules based on multinomial classification, is not widely used in marketing.

1.4   Scope of this book

Since this is an applied rather than a theoretical book, my choice of material included is based on my experience and my perception of what techniques are used frequently in marketing:

- Most attention is paid to frequently used techniques: factor analysis, correspondence analysis, cluster analysis, multiple regression analysis, discriminant analysis, and conjoint analysis.
- Related techniques such as principal components analysis and logistic regression are also covered.
- More advanced modelling techniques such as path analysis, structural equation modelling, and data mining techniques are covered, although in less detail.

I have chosen not to cover techniques that deal with multiple criterion variables and modelling techniques that are too complex and therefore not widely used in marketing. When I say not 'widely used in marketing', I mean in actual practice – in the trenches of marketing – rather in marketing literature or in academia. Journal papers tend to be an idealized version of what could be rather than a veridical version of what actually is in practice. The choice of techniques included here is necessarily arbitrary, shaped by my experience of what is widely used. I hope the choice is reasonable.

Part 2
Interdependence Techniques

2

Factor Analysis

2.1   What is factor analysis?

Marketing data tend to be voluminous. Naturally, one of the top priorities of a marketer is to summarize the data in a meaningful way. Meaningful reduction of data goes beyond simply summarizing them. It contributes to a better understanding of the underlying behavioural and attitudinal patterns of consumers.

Basics of factor analysis

The term *factor analysis* is applied to a group of techniques that transform a set of correlated variables into fewer conceptual dimensions known as 'factors'. For instance, a marketer may have measured 20 different attributes that are considered important to consumers in deciding where to shop. Based on intercorrelations among these 20 variables, factor analysis attempts to reduce them to, say, four factors such as convenience, service, price, and variety.

Factor analysis consists of two major groups of techniques: *principal components analysis* and *common factor analysis*. Briefly, principal components analysis views factors (components) as a weighted combination of variables, while factor analysis views variables as a weighted combination of factors. Factors are the *latent variables* that make up observed variables. Although both techniques provide outputs that look very similar, their underlying assumptions are very different.

However, in applied marketing research these two groups of techniques are considered to be complementary and they achieve the same goal: data reduction. As Wilkinson, Blank and Gruber (1996) pointed out: 'The differences between factors and components on real data are not large enough to justify the controversy over this issue.' In fact, principal components analysis is frequently used as the first step towards creating common factors. Following this practice, we will not make any major distinction between the two techniques.

Factor analysis is used in marketing research for three major purposes: to reduce the dimensionality of the data set; to identify the conceptual variables that underlie the measured variables; and to use as input to further statistical analysis when such analysis demands that the input variables are not highly correlated among themselves.

Consider some of the typical problems faced by marketing researchers:

1. Consumers rate a corporation on 30 different attributes that are related to customer satisfaction. It might be helpful to reduce these 30 attributes to a few (say, four or five) indices that explain most of the variance in the original 30 variables, so they can be used to summarize the information contained in the original variables.
2. A marketing manager attempts to forecast sales using a multiple regression model. The model requires that the variables used to forecast sales are not too highly correlated with each other (multicollinear). If it turns out that the variables are indeed correlated, the marketing manager may want a technique that will mathematically create new variables that are uncorrelated with each other.
3. The management of a high-technology firm would like to understand the conceptual factors that facilitate the adoption of new products fairly quickly.
4. A packaged goods manufacturer would like to isolate the basic dimensions that give rise to 30 or so evaluative variables that are being routinely used in their survey of consumers.

Problems like these can be tackled by using factor analysis.

2.2   Factor analysis model

The basic concept

To set up the basic factor analysis model we need to start with the n variables of interest – product attribute ratings, service quality measurements, or corporate image ratings, for example. We standardize these – that is, redefine them so that they have mean 0 and standard deviation 1 – and denote them $x_1, x_2, x_3, \ldots, x_n$.

The purpose of factor analysis is to express each variable as a combination of fewer underlying factors:

$$x_2 = a_{11}f_1 + a_{12}f_2 + a_{13}f_3 + a_{14}f_4 + \ldots,$$
$$x_2 = a_{21}f_1 + a_{22}f_2 + a_{23}f_3 + a_{24}f_4 + \ldots,$$
$$x_n = a_{n1}f_1 + a_{n2}f_2 + a_{n3}f_3 + a_{n4}f_4 + \ldots,$$

where the a_{ij} are weights[1] attached to different factors (f_i) for a given variable (x_i). It is unlikely that the factors we identify will explain all the variance because there might be a special characteristic associated with a given variable that is not shared by other variables (specific variance or error variance). We therefore need to add these variances into our model – we denote them e_i. If we decide that there are four factors, our factor analysis model is as follows:

$$x_1 = a_{11}f_1 + a_{12}f_2 + a_{13}f_3 + a_{14}f_4 + e_1,$$
$$x_2 = a_{21}f_1 + a_{22}f_2 + a_{23}f_3 + a_{24}f_4 + e_2,$$
$$x_3 = a_{31}f_1 + a_{32}f_2 + a_{33}f_3 + a_{34}f_4 + e_3,$$
$$x_4 = a_{41}f_1 + a_{42}f_2 + a_{43}f_3 + a_{44}f_4 + e_4.$$

The proportion of variance a variable shares with all factors combined is called the *communality* (see p. 24).

---

[1] They are in fact standardized multiple regression coefficients.

Identifying the factors

If we are to express each variable as a function of some underlying factor, we first need to identify these factors. There are many techniques available to accomplish this.

Principal components analysis

The purpose of the technique is to assign weights $a_{11}, a_{21}, a_{31}, ..., a_{n1}$ to variables x_1, $x_2, x_3, ..., x_n$ so as to maximize the variance extracted and to create the first component or factor F_1:

$$f_1 = a_{11}x_1 + a_{12}x_2 + a_{13}x_3 + ... + a_{1n} x_n,$$

subject to the constraint that

$$a_{11}^2 + a_{12}^2 + a_{13}^2 + ... + a_{1n}^2 = 1.$$

The reason for this constraint is obvious. Without it, we can increase the value of f_1 by simply increasing the value of any of the weights $a_{11}, a_{21}, a_{31}, ... a_{n1}.f_1$ is the first principal component. The variance of this factor is its *eigenvalue*.[2] $(a_1, a_2, a_3, ..., a_n)$ is called the *eigenvector*.

The second task of the technique is to create the next component f_2, uncorrelated with f_1, by creating weights $a_{21}, a_{22}, a_{23}, ..., a_{2n}$ on the variables $x_1, x_2, x_3, ..., x_n$ so as to maximize the remaining variance extracted:

$$f_2 = a_{21}x_1 + a_{22}x_2 + a_{23} x_3 + ... + a_{2n} x_n,$$

again subject to the constraint that

$$a_{21}^2 + a_{22}^2 + a_{23}^2 + ... + a_{2n}^2 = 1.$$

f_2 is the second principal component and its variance is its eigenvalue.

This process is repeated until n components $f_1, ..., f_n$ have been extracted.

A note on the common factor model

Common factor analysis, which is often used interchangeably with principal components analysis in marketing research, is a theoretical model that attempts to explain the observed correlations among variables by hypothesizing common underlying factors. Each variable is considered to be a weighted combination of a number of underlying factors. The part of the variable that is not accounted for by factors is the error term e_i (specific variance or error variance). Thus

$$x_1 = a_{11}f_1 + a_{12}f_2 + a_{13}f_3 + ... + e_1,$$
$$x_2 = a_{21}f_1 + a_{22}f_2 + a_{23}f_3 + ... + e_2,$$
$$x_n = a_{n1}f_1 + a_{n2}f_2 + a_{n3}f_3 + ... + e_n.$$

(For a brief technical description of these two basic models, see the Appendix.)

Other methods

The principal axis factors method is identical to the principal components analysis with one exception. In the diagonal cells of the correlation matrix (which contains 1.0 in principal components analysis), we insert an estimate of communality for each variable. Communality is, as noted earlier, the extent to which a given variable is

---

[2]Since factor analysis uses standardized variables, the total amount of original variance is equal to the number of variables. The amount of variance explained by each factor, its eigenvalue, is expressed in terms of the number of units of variance explained.

explained by all the factors extracted in the analysis. This is one of the most commonly used methods of extracting factors.

The minimum residual method is also similar to principal component analysis, except that the analysis is performed ignoring the diagonal entries of the correlation matrix. Factors are extracted against the criterion of minimizing the sum of squared off-diagonal residuals. This method can potentially produce communalities that are greater than 1 and can provide misleading results. If you are not well versed in factor analytic procedures, you may want to avoid this procedure.

The maximum likelihood method attempts to maximize the amount of variance explained in terms of the *population* (as opposed to the *sample,* which is the focus of principal components analysis). When communalities are high, the differences between the principal components method and the maximum likelihood method tend to be trivial. The main advantage of the maximum likelihood method is that it has statistical tests of significance for each factor extracted.

Image analysis assumes that the common core of a variable is what can be predicted by other variables. Prediction is measured by multiple regression analysis (see Chapter 7) of each variable against all other variables. These analyses produce a covariance matrix which, in turn, is subjected to factor analysis. The results produced through this analysis are somewhat difficult to understand and interpret. This method tends to give results that are different from other methods of factoring.

Alpha factor analysis attempts to create factors that have the highest reliability. Reliability refers to internal consistency and consistency over time. No significant advantage is attached to this method compared to other methods of factor extraction.

Choosing a method for factoring

Principal components analysis is useful when we want to explain the maximum amount of variance in our data with the minimum number of factors (or components, as they are called in principal components analysis). It is also a technique to be used when we want to use the resulting factor scores in subsequent analyses (such as input to regression analysis).

Principal axis factoring is also fairly commonly used. It is useful when we want to identify the underlying dimensions and when common variance is our focus.

While all the factoring methods are used by researchers, most tend to use either principal components analysis or principal axis factoring. Many analysts prefer to use principal components exclusively because it is based on fewer assumptions than most other techniques.

Number of factors

To achieve data reduction, the number of factors extracted should be (far) less than the number of variables. No data reduction is achieved if we extract 20 factors out of 20 variables. In fact, this is exactly what happens when we extract factors using principal components – we can create as many factors as there are variables. So it is up to the analyst to decide how many factors are acceptable. Here are some common criteria used to guide the analyst.

Variance explained

Some researchers accept as many factors as would explain a predetermined amount of variance, say 70%. This is one of the weakest ways of deciding the number of factors, since here our prior decision rather than the patterns in the data determine the number of factors.

Scree criterion

To use the scree criterion, plot the eigenvalue of each component on a graph and connect the points (see Exhibit 2.1). We accept all factors until the slope of the curve starts changing (the 'elbow effect'). The idea behind the scree criterion is that random variations are seen mostly at the tail end of the curve. The problem with this procedure is that there may be *no* abrupt change in the curve.

EXHIBIT 2.1

Using the scree criterion

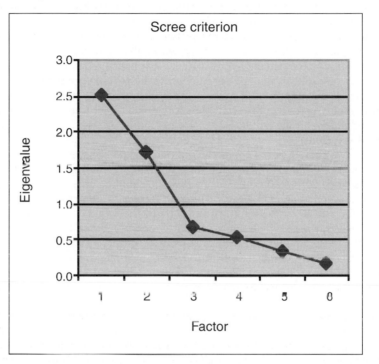

Eigenvalue criterion

This is perhaps the most widely used criterion – we accept all factors that have an eigenvalue of 1 or more. The logic behind this approach is that it is intrinsically less appealing to create a weighted combination of variables to explain less variance than a variable would explain on its own. Since the variables are standardized, this criterion will result in a conservative number of factors.

Sometimes the scree criterion is used to further reduce the number of factors identified by the eigenvalue criterion.

Factor matrix

A major result of factor analysis is the *factor matrix* (also called the factor pattern matrix). It contains *factor loadings*, or the correlations between factors and variables. Factor loadings are given by:

$$\frac{w_{ij}}{s_j} \sqrt{\lambda_i},$$

where w_{ij} is the weight for jth variable on the ith factor, s_j is the standard deviation of the jth variable and λ_i is the eigenvalue of the ith factor.

A high absolute coefficient indicates that the variable and the factor are highly related. Since we do not know beforehand what a factor stands for, we use loadings to assign meaning to it. For example, if a factor is highly loaded on variables such as discount, low cost, and free after-sales service, we can conclude that this factor is price since that is what is common to all highly loaded variables.

Factor rotation

In practice, the initial factor matrix is difficult to interpret because factors tend to be correlated with many variables (which may not even have a common theme). To make the factors interpretable, we redistribute the variance explained by each factor through a procedure known as *factor rotation*. Rotation does not affect either the communalities or the percentage of variance explained. Rotation simply minimizes the number of variables that are strongly correlated with a given factor, while at the same time maximizing the size of the correlations of a few variables. When each factor loads highly on only a few variables, it is much easier to understand the conceptual underpinning of the factor.

Types of rotations

There are two basic types of rotations: orthogonal and oblique.

In *orthogonal rotation*, factors are rotated in such a way that they will continue to be uncorrelated with one another, even after the rotation. This means that the factors are rotated at right angles (90°) to each other. *Varimax rotation* is an orthogonal rotation method and is by far the most popular. In the varimax procedure, factors are rotated such that each variable loads highly on one factor and only one factor. Thus each factor will represent a distinct construct. In *quartimax rotation,* factors are rotated such that all variables will be loaded highly on one factor. In addition, each variable will have a high loading on one other factor but near-zero loadings on the remaining factors. This results in one overall factor followed by a number of specific factors.

In *oblique rotation*, the constraint – that factors be rotated such that they are at right angles to each other – does not exist. We can rotate the factors such that they are at any angle to each other and therefore the factors can be correlated with each other. *Quartimin, oblimin, covarimin, binormamin, orthoblique, doblimin, dquart,* and *promax* are some of the oblique rotation procedures.

Choosing a method of rotation

Oblique rotation is appealing to some users because it is less constrained. However, oblique solutions are less widely used than orthogonal ones, possibly because of the potential problems of interpretation. When we use oblique solutions, the sum of squared loadings in each row does not equal the communality (h^2), except by chance. As a result, we cannot clearly determine the proportion of variance in a variable explained by the factors. Neither can we determine the variance explained by each factor. So, it is generally preferable to use an orthogonal method such as varimax. Analysts who choose to use an oblique solution should weigh the statistical losses against possible gains in meaning.

Factor scores

Since factor analysis reduces a large number of variables to fewer factors, if we choose to, we can replace the scores on individual variables with factor scores. For example, if we identify a price factor that is highly loaded on variables such as discount, low cost, and no-cost after-sales service, we can replace the individual ratings on these variables with a single factor score for price. The factor score F_i is given by

$$F_i = w_{11}x_1 + w_{12}x_2 + w_{13}x_3 + \ldots + w_{1n}x_n.$$

Such factor scores may be used as input to subsequent analysis. Analysts who are uncomfortable using conceptual variables rather than measured variables often use a single variable to represent a factor. The chosen variable is generally the highest-loaded variable for each factor. These variables are called *principal variables*.

Model fit assessment

Model fit is assessed in many ways. The first concern of the analyst is whether the analysis makes sense. Are the factors logical? Are they interpretable? Are they usable? Once the factors are judged to make sense, the second concern is to assess the total amount of variance explained by the extracted factors. There is no hard-and-fast rule here but, in general, the greater the amount of variance that is explained the better the fit. Finally, because factor analysis assumes that the observed correlation between variables can be attributed to common factors, it follows that correlations can be deduced from factor loadings. The difference between observed and deduced correlations (residuals) shows how well the model fits.

2.3 Factor analysis: computer output

A financial services firm surveyed 1000 people and asked them to rate the importance of six different variables – price, product quality, discount, reputation, well-established, and service quality – that are important to them in choosing a financial services firm. Factor analysis (principal components analysis with varimax rotation) was applied to the data to summarize most of the information contained in the six variables into fewer dimensions. The computer output (Systat) is shown in Exhibit 2.2.

The starting point for all factor-analytic techniques is the the correlation matrix (shown in the output). Since principal components analysis creates as many factors as there are variables, six factors are created, along with their eigenvalues. If we use the eigenvalue criterion to determine the number of factors, we only have two factors whose eigenvalues exceed 1.0. We retain these two factors for further analysis. These two factors together explain about 70% of the variance (42.22% + 28.72%), as shown in the output under 'Percent of total variance explained'.

To interpret these factors, we examine the component (factor) loadings that follow. However, as is typical, we note that the first factor loads highly on most variables, making interpretation difficult. When the factors were rotated using the varimax procedure ('Rotated loading matrix'), the patterns become clearer: the first factor loads highly on four variables, and the second factor on the remaining two. The variables are rearranged to make this pattern obvious. Although the individual loadings have changed, the total variance explained by the rotated factors remains the same – about 70% (38.87% + 32.08%).

To understand the meaning of the two factors, we need to examine the 'Rotated loading matrix' more closely. More specifically, we examine those variables that load highly on each factor. If a given factor loads on some variables then we look for the common theme that unites those variables and thus interpret what the factor means. We can name the factor accordingly.

EXHIBIT 2.2

Factor analysis output (principal components analysis), annotated

Correlation matrix

	PRICE	PROQLTY	DISCOUNT	REPUTATN	ESTABLSHD	SERVQLY
PRICE	1.00					
PRODUCTQLTY	−0.29	1.00				
DISCOUNT	0.48	−0.51	1.00			
REPUTATION	0.02	0.29	−0.08	1.00		
ESTABLISHED	0.05	0.20	−0.04	0.82	1.00	
SERVICEQLTY	−0.52	0.50	−0.39	0.23	0.21	1.00

Component loadings (*Same as factor loadings. Refer to correlations between a factor and each variable in the analysis*)

	1	2	3	4	5	6
PRICE	−0.59	0.53	−0.47	0.23	0.30	−0.02
PRODUCTQLTY	0.76	−0.11	−0.49	0.32	−0.27	0.05
DISCOUNT	−0.68	0.41	0.34	0.44	−0.26	0.02
REPUTATION	0.55	0.77	0.04	−0.10	−0.07	−0.29
ESTABLISHED	0.50	0.80	0.11	−0.13	0.04	0.20
SERVICEQLTY	0.77	−0.20	0.31	0.41	0.31	−0.03

Latent Roots (Eigenvalues)

1	2	3	4	5	6
2.53	1.72	0.68	0.55	0.33	0.17

The six eigenvalues above correspond to the six factors (not to the measured variables). Only the first two factors have an eigenvalue greater that 1. We choose to retain these two factors.

Percent of Total Variance Explained (*Eignevalues divided by the number of variables × 100*)

1	2	3	4	5	6
42.22	28.72	11.41	9.18	5.58	2.88

Component loadings (*factor loadings*)

	1	2
SERVICEQLTY	0.77	−0.20
PRODUCTQLTY	0.76	−0.11
DISCOUNT	−0.68	0.41
PRICE	−0.59	0.53
REPUTATION	0.55	0.77
ESTABLISHED	0.50	0.80

Note how every variable loads highly on the first factor.

Rotated Loading Matrix (VARIMAX)

	1	2
DISCOUNT	−0.79	0.01
PRICE	−0.78	0.16
SERVICEQLTY	0.77	0.21
PRODUCTQLTY	0.71	0.29
REPUTATION	0.10	0.94
ESTABLISHED	0.03	0.94

Varimax rotation has clarified the meaning of factors by readjusting the loadings such that some variables load highy on factor 1 while other variables load highly on factor 2.

'Variance' Explained by Rotated Components

Eigenvalues of the first 2 factors ⟶

	1	2
	2.33	1.92

Percent of Variance Explained

	1	2
	38.87	32.08

Although the explained variance has been redistributed, note that the total variance explained by the components has remained the same.

Latent Vectors (Eigenvectors) *These are the weights assigned to each variable in the creation of factors. For instance, Price contributes negatively to factor 1 (-0.37) and positively to factor 2 (0.40).*

	1	2
PRICE	−0.37	0.40
PRODUCTQLTY	0.48	−0.08
DISCOUNT	−0.42	0.31
REPUTATION	0.35	0.59
ESTABLISHED	0.31	0.61
SERVICEQLTY	0.48	−0.15

Factor 1 loads high on price, discount, product quality, and service quality. Since these four variables now form a clear group, the factor can now be named, perhaps as 'cost vs. quality' (rather than 'general liking' as we would have named if it loaded highly on every variable, as it did in the unrotated matrix). The second factor loads highly on reputation and 'established'. This factor can be named as 'Image'.

Unipolar and bipolar factors

For factor 1, some high loadings were positive (product and service quality) while other loadings were negative (price and discount). Such factors are called *bipolar* factors. In our example, our first factor is cost vs. quality, meaning quality and price are perceived to be the opposite ends of the same dimension, as opposed to their being two distinct dimensions. When all the high loadings within a factor are in the same direction (all positive or all negative), we have a *unipolar* factor. Factor 2, in which both high loading variables are positive, is an example of a unipolar factor.

Mapping the variables

The relationship between the factors and variables can also be represented visually. The map at the end of Exhibit 2.2 shows how different variables are represented in relation to the first two factors. While this is not of much interest to us at this stage, we may want to note that such visual representation is possible. When we discuss techniques such as correspondence analysis and biplots, the usefulness of such representations will become obvious. These techniques use principal components analysis in producing perceptual maps.

Communality

It is possible, and is quite common, that some variables do not relate to any of the factors being created. This is of concern to us, since it limits our ability to use factors to represent all variables. To assess how well a variable is represented by a factor, we examine the communality of that variable. Communality is calculated by summing the squared loadings across all factors. Thus the communality for first variable in our example (discount) is $0.79^2 + 0.01^2 = 0.64$, or 64% of the variance of the variable has been explained by the two factors under consideration.

If the communality of a variable is too small, say 0.3 or lower, and if the variable is considered important from a marketing perspective, we may need to pay special attention to that variable, so it can be accommodated separately in our future analyses if necessary.

Factor scores

Since we have chosen to replace all the variables in the original study with the two factors, we may now replace the original ratings with factor scores. You need to assign different weights to different variables depending on how much each variable contributed to the make-up of a component. This can be discerned from the 'Latent vectors or eigenvectors' table in Exhibit 2.2. It shows what weights were assigned to each variable to arrive at each factor. For instance, factor 1 is a weighted combination of variables in which price was assigned a weight of –0.37, product quality was assigned a weight of 0.48 and so on. To arrive at the factor score for an individual for a given factor, we multiply that individual's ratings on all variables by the corresponding weight for that factor and sum them. Computer programs can automatically calculate the factor scores for each respondent.

Once we have the factor score for each respondent, these scores can replace the original variables and can be used in further analysis. For instance, we may want to predict how each factor influences purchase behaviour, using regression analysis (see Chapter 5). Or we may want to predict how factor scores of social variables influence voting behaviour, using discriminant analysis (see Chapter 6). While these types of analyses are commonly done, you should be careful if the factors are less than clear-cut or if the factors fail to explain a large amount of variance in the data.

2.4 Factor analysis: marketing applications

In this section, I present three applications of factor analysis to marketing problems. They illustrate the variety of seemingly unrelated problems to which factor analysis may be applied.

Is shopping related to the specific retail environment?

Marketing problem
In the United Kingdom, unlike in the United States, high street shopping represents a traditional retail location. Shopping malls are a relatively recent phenomenon. High street shopping is comparable to mall shopping in terms of area, variety and mix. Since both shopping units compete for business, it would be useful to understand how consumers view these two alternatives.

Are there significant differences between the mall and the high street shopping area in terms of consumer experience and behaviour? Hackett and Foxall (1994) addressed this issue through a research project.

Application of factor analysis
Hackett and Foxall chose Merry Hill shopping mall and Worcester city centre to re-present malls and high street shopping areas. Interviewers who were placed in these two locations approached shoppers and handed them a questionnaire (identical for both locations) that contained 33 attributes that were potentially important to shoppers. (These 33 items were: staying dry, high-quality goods, value for money, helpful staff, clean shopping areas, a wide choice of goods, shops being close to your home, staying warm, good road links, small specialist stores, convenient parking, being able to easily find the shops you are looking for, friendly staff, places to sit down, easy access to information about shops and the goods they offer, cafés, an attractive-looking shopping area, toilets, a wide choice of shops, large supermarkets/hypermarkets, a baby feeding/changing area, good public transport links, places to meet others, shopping with friends, getting a bargain, enjoying yourself, large department stores, litter-free shopping areas, a place to leave children, security staff, short distances between shops, pedestrian zones, and public houses.)

Shoppers were asked to rate the importance of each of these 33 items on a four-point scale that ranged from very important (1) to very unimportant (4). In addition, shoppers were also asked to provide demographic information. In all, 204 shoppers were interviewed (101 in Merry Hill and 103 in Worcester city centre).

Hackett and Foxall analysed the data using factor analysis for each location independently. Their hypothesis was that if different factors influenced the shoppers to shop in these two different locations, then the result of factor analysis would reflect this.

Results of factor analysis
The researchers applied principal components analysis with varimax rotation to both sets of data (corresponding to the two locations).

Merry Hill. The Merry Hill data identified 11 factors using the 'eigenvalue greater than 1' criterion. To effect further data reduction, the analysts created a scree plot of eigenvalues. The plot indicated that the impact of the first three factors was far greater than those of the remaining factors. In terms of variance, the first three factors together explained 38%, of the variance while the remaining eight factors taken together explained 34% of the variance. The factor loadings of the first three factors are given in Exhibit 2.3. (Only variables that have loadings equal to or greater than 0.4 are included in the exhibits.)

EXHIBIT 2.3

Factor loadings for Merry Hill

	Loading*
Factor 1. Store variety	
A wide choice of shops	0.80
A wide choice of goods	0.78
Large departmental stores	0.63
Being able to easily find the shops you are looking for	0.62
Getting a bargain	0.44
Factor 2. Comfort and convenience	
Staying dry	0.81
Staying warm	0.77
Shops being close to your home	0.52
Easy access to information	0.51
Small specialist stores	–0.40
Factor 3. General facilities and convenience	
A baby feeding/changing area	0.70
Security staff	0.62
Large supermarkets/hypermarkets	0.61
A place to leave children	0.56

*Loadings in this table are correlations between factors and each of the variables that load highly on that factor.

Worcester city centre. The Worcester city centre data also identified 11 factors using the eigenvalue criterion. As before, the analysts created a scree plot of eigenvalues that indicated that only four factors were strong enough to meet the scree criterion. In terms of variance, the first four factors together explained 41% of the variance, while the remaining seven factors taken together explained 29% of the variance. The factor loadings of the first three factors are given in Exhibit 2.4.

How factor analysis addressed the problem

The first and obvious conclusion is that the factors of attribute importance are different for these two locations. This means that, presumably, shopping preference is related to what shoppers consider important to them. For mall shoppers it is *store variety, comfort and convenience,* and *general facilities and convenience*, and for

EXHIBIT 2.4

Factor loadings for Worcester city centre

	Loadings
Factor 1. Service quality	
Helpful staff	0.89
Friendly staff	0.83
Clean shopping areas	0.70
High-quality goods	0.62
Value for money	0.58
Enjoying yourself	0.43
Factor 2. Access and facilities	
A baby feeding/changing area	0.85
A place to leave children	0.79
Security staff	0.64
Good road links	0.47
Factor 3. Social	
Places to meet others	0.84
Shopping with friends	0.73
Public houses	0.70
Places to sit down	0.49
Good public transport links	0.41
Factor 4. Choice and variety	
A wide choice of goods	0.78
A wide choice of shops	0.68
Being able to easily find the shops you are looking for	0.58
Short distances between shops	−0.54
Pedestrian zones	0.43

high street shoppers it is *service quality, access and facilities, social,* and *choice and variety*. Since the questionnaire was common for both locations, the differences in factors cannot be attributed to the questionnaire. (Since factors are dependent on the questions asked, changes in the questions could result in different factors.)

The second conclusion is that certain items may be common to both shopping environments while others may be context-related. For instance, the first factor at Merry Hill (store variety) resembles the fourth factor at Worcester (choice and variety). In each case, the factor absorbs some supplementary items that are specific to that location.

Of particular interest are factors that appear in one context but not in another. As an example, when we examine factor 2 at Merry Hill (comfort and convenience), we note that this factor does not appear at all at Worcester. None of the variables that make up the factor show up as a part of any Worcester factor. Similarly, factor 3 at Worcester (social) does not show up in any form at Merry Hill. This was interpreted to mean that there might be distinct motivations that contribute to mall shopping as opposed to high street shopping.

Based on these results, the researchers came to the following conclusions:

1. The motivations of those who shop at the mall are not the same as those who shop on the high street.
2. The same attribute can be a part of a different conceptual package (factor) depending on the chosen shopping environment.
3. Research methodology that assumes that shopping motives are common for all shoppers and attempts to compare the two shopping preferences in terms of standard attributes such as distance from home, price and variety may provide misleading information.

Do different brands have different personalities?

Marketing problem

By branding a product, a company distinguishes its products from those made by others. Many companies may make cars, but Ford is not the same as General Motors. Even among the cars made by the same company – such as the Ford Escort and Ford Mustang – the products are distinguished through branding. A brand is considered to be more than a nametag. It is more than a summary of product features. Many studies suggest that brands – like humans – have personalities attributed to them by consumers. For instance, Virginia Slims cigarettes tend to be considered 'feminine' and Marlboro 'masculine' (Aaker, 1996, p. 142). As Walter Landor (Meyers and Lubliner, 1998) puts it: 'Products are created in a factory, but brands are created in the mind.' If we accept the proposition that brands have distinct personalities, it is important for marketers to understand the different brand personalities so they can effectively market the product to different consumer segments. But how do we measure the personalities of different brands within a product category? How do we compare brand personalities across different product categories? Personality research studies carried out by psychologists have converged to identify a reliable factorial composition of five underlying personality traits known as the 'Big Five'. How applicable are they to brands?

Application of factor analysis

In order to develop a theoretical framework of brand personality dimensions, Jennifer Aaker (1997) applied factor-analytic techniques to specially collected survey data. Aaker's research centred around the 'Big Five'. How applicable are they to brands?

Aaker sent out a predesigned questionnaire to 1200 members of a national mail panel. Of these, 631 (55%) responded to the questionnaire. The respondents were asked to rate 37 national brands on 114 personality traits. Consumers were asked to rate each product on each personality trait on a seven-point scale in which 1 stood for 'not at all descriptive' and 7 stood for 'extremely descriptive'. Factor analysis was applied to these data.

Results of factor analysis

To apply factor analysis, Aaker correlated the 114 attributes with one another, thus creating an attribute correlation matrix of order 114 × 114 across all 37 brands. She then subjected the matrix to principal component factor analysis followed by varimax rotation. The scree criterion was applied to factors whose eigenvalues exceeded 1, which resulted in the identification of five basic factors (see Exhibit 2.5). These five

EXHIBIT 2.5

Five brand personality factors

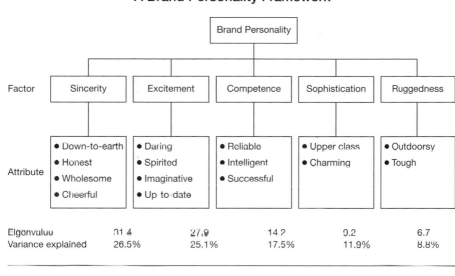

A Brand Personality Framework

	Sincerity	Excitement	Competence	Sophistication	Ruggedness
Eigenvalue	21.4	27.9	14.2	9.2	6.7
Variance explained	26.5%	25.1%	17.5%	11.9%	8.8%

factors were judged to be robust for a number of reasons:

1. Each of the five factors loaded highly on a distinct set of variables (but not the remaining variables).
2. The five factors together explained 92% of the variance of attribute ratings, which can be considered a very high degree of explanation.
3. The same five factors recurred when it was applied to different subsamples such as men, women, younger respondents and older respondents.
4. Another sample of 180 respondents rated a different set of brands on 42 attributes. A confirmatory factor analysis (a procedure not explained in this chapter) applied to the data confirmed the five-factor model.

Exhibit 2.6 shows the scores of four selected brands on each of the five factors.

EXHIBIT 2.6

Brand personality scores

	Computers		Broadcasters	
	Apple	IBM	CNN	ESPN
Sincerity	0.92	0.89	0.99	0.99
Excitement	0.95	0.91	1.02	1.10
Competence	1.07	1.10	1.18	1.04
Sophistication	0.86	0.84	0.93	0.89
Ruggedness	0.92	0.91	1.01	1.23

When we compare the brand personality scores of two computers and two broadcasters, some patterns become apparent. Apple and IBM make computers that are very different. Similarly, CNN's programming (news-oriented) is very different from that of ESPN (sports-oriented). Yet, we note that there is greater similarity of brand personality between brands within each category than between them. Compared to CNN, ESPN is seen to be more 'rugged', which presumably relates to its physical activity-oriented programming. From the results (not shown above) it also would seem that different product categories have different personalities – none of the cars in the study scored high on sincerity while toy manufacturers like Lego and Mattel are rated high on all factors except sophistication.

How factor analysis addressed the problem

The application of factor analysis established that it is possible to develop a reliable framework for measuring brand personality. This was confirmed by the high proportion of explained variance, subgroup analysis and the results of confirmatory factor analysis.

Second, factor analysis uncovered five underlying dimensions of brand personality: sincerity, excitement, competence, sophistication and ruggedness.

Third, the analysis compared and contrasted the similarities and differences between human personality and brand personality. Exhibit 2.7 shows the 'Big Five' human personality factors and brand personality factors identified by Aaker.

EXHIBIT 2.7

Human and brand personality factors

Human personality factors	Brand personality (approximate equivalent) factors
Adjustment	No equivalence
Agreeableness	Sincerity
Extroversion	Excitement
Conscientiousness	Competence
Imagination/curiosity	No equivalence
No equivalence	Sophistication
No equivalence	Ruggedness

In summary, the application of factor analysis to the brand personality problem identified the basic factors of brand personality. As a result, it provided a means of comparing different brands and, by inference, possibly the images of product categories, and showed the similarities and differences between human and brand personality factors.

Is liking a product related to its sensory attributes?

Marketing problem

One of the challenges faced by marketers is the task of describing a product. While every product has certain physical and chemical properties, marketers are less concerned about such characteristics. Rather, they are interested in how consumers perceive the product characteristics. Descriptions such as 'pleasant', 'unpleasant' and 'refreshing' that consumers use to describe a product may or may not have any relationship to the objective characteristics of the product itself. To complicate this further, any positive attributes that a consumer assigns to a product may simply be the result of his or her liking a product: positive characteristics are attributed to the liked brand and negative characteristics are attributed to the disliked brand. Sensory attributes of a product may comingle for certain products but not for other products. Given the importance of consumer language and perception in marketing communications, it becomes critical to understand the relationship between product attribute evaluations and liking. How do different product attributes comingle with overall liking for that product? The research cited here studies this marketing problem in relation to two products: meat and carbonated drinks.

Application of factor analysis

As a part of an attempt to understand the relationships among liking, sensory, and directional attributes of different product categories, Moskowitz (1999) tested two product categories: meat and carbonated drinks.

In the meat study, 150 panelists evaluated different meat products. The panelists were poultry consumers drawn from four US markets and rated the products on a variety of liking, sensory, directional, and image attributes. The scale value ranged from 0 to 100.

In the carbonated drinks study, 200 panelists evaluated different carbonated drinks. The panelists were soft drink users. Using a methodology similar to that used in the meat study, consumers evaluated soft drinks on a variety of liking, sensory, directional, and image attributes. Again, the scale value ranged from 0 to 100.

Moskowitz applied factor analysis to these two sets of data to assess the relationship between sensory attributes and liking in these two product categories.

Results of factor analysis

For each product category, factor analysis was applied to attribute ratings that also included the overall liking rating for the product under consideration. Principal components analysis followed by quartimax rotation was applied in both instances. Factors with an eigenvalue in excess of 1 were accepted for further interpretation and use. This yielded four factors for meat products that together explained 84% of the variance of the attribute ratings. For carbonated drinks, factor analysis yielded three factors that together explained 83% of the variance. The factor loadings for these two products are presented in Exhibits 2.8 and 2.9.

The difference between the two product categories is quite striking. For carbonated drinks, overall liking is part of the first and the strongest factor that includes a number of sensory attributes such as flavour, strength, after-taste, carbonation and sweetness. The second and third factors deal with narrowly specified attributes such as aroma, and citrus flavour.

For meat, on the other hand, the first factor that comprised sensory attributes did not show a strong relationship to liking. (In fact, overall liking shows a mild negative relationship to sensory attributes.) Rather, overall liking shows up as the third factor with only one related variable: 'meaty'.

How factor analysis addressed the problem

The results helped to clarify the relationship between liking and sensory attribute evaluation for the two product categories. In fact, an examination of the factor structure of these two products leads us to a number of clear conclusions related to our marketing problem.

EXHIBIT 2.8

Factor structure for meats: sensory attributes and liking

Attributes	Factor 1	Factor 2	Factor 3	Factor 4
Flavour intensity	**0.92**	0.24	0.15	−0.15
Aroma intensity	**0.90**	−0.09	−0.01	0.05
Aftertaste	**0.85**	0.32	0.04	−0.06
Spicy	**0.82**	0.36	0.27	0.00
Red	**0.80**	−0.53	−0.12	0.09
Sweet	**0.79**	0.34	0.03	0.24
Dark	**0.77**	−0.56	−0.15	0.07
Greasy	**0.75**	0.58	−0.02	0.02
Salty	**0.69**	0.51	0.25	−0.16
Juicy	0.59	**0.72**	0.05	−0.03
Firm	−0.09	**−0.86**	0.32	0.28
Thick	0.48	**−0.64**	−0.08	−0.46
Meaty	0.28	−0.54	**0.62**	−0.42
Liking	−0.43	0.01	**0.83**	−0.04
Gritty	0.52	−0.32	0.01	**0.70**
Even appearance	0.49	−0.09	0.08	0.26
% Variance explained	45.50	21.44	10.62	6.51

EXHIBIT 2.9

Factor structure for fruit-flavoured soft drink

Attributes	Factor 1	Factor 2	Factor 3
Tart flavour	**0.96**	0.06	0.13
Harsh flavour	**−0.91**	−0.01	0.26
Flavour strength	**0.80**	0.38	0.16
Overall liking	**0.79**	0.05	0.54
Strength of aftertaste	**−0.79**	0.41	−0.02
Carbonation	**0.74**	0.15	0.02
Sweetness	**−0.64**	0.27	0.10
Lime aroma	0.16	**0.88**	0.22
Aroma strength	0.42	**0.88**	−0.07
Lemon aroma	0.06	**0.87**	0.28
Lime flavour	−0.10	0.24	**0.94**
Lemon flavour	−0.18	0.21	**0.92**
% Variance explained	40%	24%	19%

1. Each product is evaluated on several factors.
2. The dynamics of these factors depend on the product category.
3. Overall liking plays a very small role in the factor structure for meats. Being 'meaty' rather than sensory attributes relates to overall liking for meat.

4. For carbonated drinks, consumers do not easily separate sensory attributes from liking. Hedonistic evaluation is pervasive to carbonated soft drink products.
5. Comingling of sensory and liking attributes varies by product category.

For a marketer, patterns like these can be of considerable importance in planning marketing campaigns that aim to reach consumers. The main result of the factor analyses reported here – that sensory attributes are inextricably linked to liking for carbonated soft drinks but not for meats – is not an intrinsically obvious finding.

2.5 Caveats and concluding comments

While performing factor analysis it may be helpful to keep in mind some of the common ways in which results are misinterpreted.

Interpreting the factors without regard to the correlations

Factor analysis will generate factors even when the correlations are low (Chakrapani and Ehrenberg, 1980, 1981). Since the analysis is based on the relative rather than the absolute sizes of correlations, unless we examine the correlation matrix to make sure that the correlations are large enough to be meaningful, our interpretation may be far removed from reality.

Treating factors that explain more variance as 'important'

By the way the analysis is structured, the first factor will explain the largest amount of variance. Subsequent factors will explain less and less variance. For two reasons, the factor that explains the largest amount of variance may be no more 'important' than one that explains much less variance. First, factor analysis is an interdependent (grouping) technique. We have not tested the factors against any other criterion that will enable us to make a statement with regard to their importance. Secondly, the variance explained simply depends on how many highly correlated variables there were in the original data. If our original data had a number of related questions about product features but only two questions on pricing, product features would explain more variance than pricing. But that may only be because we asked only two questions about pricing and many highly correlated questions about product features. Variance explained in this context refers to the *amount* of information contained in the original set and not to its *importance*. Unless we have additional information available, a factor that explains 60% of the variance has no more marketing significance than the one that explains only 10% of the variance. Similarly a factor analysis solution that explains 90% of the variance may contain no relevant marketing information.

Ignoring variables that do not load highly on any factor

This is related to the previous mistake. If a variable does not load on any factor significantly, it simply means that the extracted factors were not effective in summarizing the information contained in that variable. It does *not* mean that the variable itself is not 'important'. We should pay particular attention to variables that do not load on any factor significantly. (The variables will have a low communality value.) Suppose we ask 30 questions about product features and only one question about pricing. Price may not load on any component, even though it may be a critical variable in decision-making.

Confusing the label with the content

Another limitation that relates to the use of factor analysis has to do with the naming of factors. Since most users of applied research tend not to look too closely at the data, careless labelling of factors can lead to misleading interpretations. As an example, a food group factor that includes food items that are ready to eat can conceivably be named either 'convenience food' or 'junk food'. Once the factor is named, the marketing and communications strategy that follows will be geared more to the label than to the contents of the factor. An advertising campaign developed around the 'junk food' theme could be very different from a campaign developed around the 'convenience food' theme. Yet it may simply be an artefact, dependent solely on how a factor was named by the analyst.

Factor analysis is one of the most frequently used methods in marketing and research. Although its main use has been data reduction, as our examples show, factor analysis can be applied to a variety of problems. While the application of factor analysis for data reduction purposes mostly confirms our intuitive ideas, this is not necessarily so when it is applied to problems that deal with 'pattern recognition' of consumer thought processes, as illustrated in the research studies outlined in the previous section. It is for this reason that the use of factor analysis has been finding increasing use in marketing.

Bibliography

Further reading

For a detailed exposition of factor analysis readers are referred to:
Jackson, J. Edward (1991) *A User's Guide to Principal Components*. Wiley Interscience, New York.
Less technically oriented readers will benefit from:
Kline, Paul (1994) *An Easy Guide to Factor Analysis*. Routledge, London.

References

Aaker, David (1996) *Building Strong Brands*. Free Press, New York.
Aaker, Jennifer (1997) Dimensions of measuring brand personality. *Journal of Marketing Research*, **34** (August), 347–56.
Chakrapani, C. and Ehrenberg, A.S.C. (1980) An alternative to factor analysis in marketing research – Part 1: The problems of factor analysis. *Journal of the Professional Marketing Research Society,* **1**(1), 22–30.
Chakrapani, C. and Ehrenberg, A.S.C. (1981) An alternative to factor analysis in marketing research – Part 2: Between group analysis. *Journal of the Professional Marketing Research Society,* **1**(2), 32–8.
Hackett, Paul M. and Foxhall, Gordon R. (1994) A factor analytic study of consumers' location specific values: A traditional high street and a modern shopping mall. In G.J. Hooley and M.K. Hussey, *Quantitative Methods in Marketing*. Dryden Press, London.
Meyers, Herbert M. and Lubliner, Murray J. (1998) *The Marketer's Guide to Successful Package Design*. NTC Business Books, Chicago.
Moskowitz, Howard (1999) Improving the 'actionability' of product tests: Understanding and using relations among liking, sensory and directional attributes. *Canadian Journal of Marketing Research*, **18**, 31–45.
Wilkinson, Leland, Blank, Grant and Gruber, Christian (1996) *Desktop Analysis with Systat*. Prentice Hall, Upper Saddle River, NJ.

3

Correspondence Analysis

3.1 What is correspondence analysis?

Correspondence analysis is used in marketing to create perceptual maps. Typically, perceptual maps are visual representations of cross-tabulated non-metric data sets such as different brands and the demographic traits of those who use those brands. Correspondence analysis shows the structure of the data, which then lends itself to a further and more detailed analysis of data. So it is both a means of data analysis and a tool for communication.

The bulk of market research data comes in the form of cross-tabulated data, technically known as contingency tables. A typical contingency table may show the demographic profiles of users of a number of different brands. A table in which the cells contain the frequencies of association between the categories of two variables is also a contingency table. In analysing such a table, a number of questions arise: How close are two brands in terms of their user profiles? How close are certain demographic traits in relation to brands? Knowing the answer to the first question will help us identify a set of brands that might be competing for the same market niche. Knowing the answer to the second question may provide some clues as to the nature and characteristics of the segment with relation to the brands under consideration.

Such relationships among any two non-metric variables – brands and demographics, products and perceived attributes of products, magazines and reader profiles, corporations and image attributes – can be modelled using correspondence analysis.

Here are some problems for which correspondence analysis may be applied:

1. Through research, a firm identifies 10 key attributes (such as 'makes high-quality products', 'seldom makes billing errors') that lead to customer loyalty. During the next phase, the marketer carries out a study to identify how different corporations are perceived on these 10 key attributes. Customers are asked to identify the presence or absence of 10 key attributes that a corporation may possess. Based on this data set, the marketer may create a perceptual map to visually depict which corporations are perceived to be close in the way they are perceived in terms of these key attributes and which of these attributes are interrelated.

2. A magazine publisher would like to have a better understanding of the way

readership profiles have changed over the years due to the increased availability of other information sources such as the Internet. The publisher would like to relate the profiles of readers as they existed for each of the past 10 years.
3.   A car manufacturer would like to relate different types of car (such as the SUV, the sports car, and the compact car) to the characteristics of their buyers.
4.   A tour operator would like to understand how different destinations are related to destination preferences by the demographic characteristics of tourists (i.e., age group and income level).

A typical contingency table to which correspondence analysis may be applied is given in Exhibit 3.1. In this exhibit, the figure at the intersection of an attitude and a brand stands for the number of times that attitude was found to be associated with that brand in this particular survey. Correspondence analysis answers question such as: Which cereals are perceived similarly? Which attributes are closely related to one another? How do attributes relate to specific brands?

EXHIBIT 3.1

Cereals and attributes

	Corn Flakes	Weetabix	Rice Krispies	Shredded Wheat	Sugar Puffs	Special K	Frosties	All Bran
Comes back to	65	31	10	10	5	6	6	7
Tastes nice	64	40	32	23	29	17	22	11
Popular with all the family	59	30	20	13	15	7	13	5
Very easy to digest	60	42	24	18	20	20	19	17
Nourishing	40	50	17	31	18	19	14	19
Natural flavour	47	39	11	28	6	15	5	18
Reasonably priced	60	37	9	12	5	6	5	6
A lot of food value	27	38	9	26	9	17	7	17
Stays crispy in milk	42	6	23	18	13	8	11	6
Helps to keep you fit	24	28	10	21	9	29	9	40
Fun for children to eat	17	12	57	5	50	5	43	0

3.2 Correspondence analysis model

Correspondence analysis terminology

Correspondence analysis in its current form was developed and popularized by French data analysts such as Jean-Paul Benzécri. During the course of its development, the developers of the technique used their own special terminology, which is commonly used in the literature, to describe well-known concepts. It is important to be familiar with the terminology not only to follow published research in the field but also to use some of the correspondence analysis software which may describe the results in terms of this terminology.

Profiles. Profiles are proportions or relative frequencies, obtained by dividing a set of frequencies (a non-negative quantity) by their sum, as shown in Exhibit 3.2.

EXHIBIT 3.2

Row and column profiles

Raw data

	High income	Low income	Total
Men	100	300	400
Women	100	100	200
Total	200	400	600

Row profiles (gender profiles across income)

	High income	Low income
Men	0.25	0.75
Women	0.50	0.50
Profile of marginal row	0.33	0.67

Column profiles (income profiles across gender)

	High income	Low income	Profile of marginal column
Men	0.50	0.75	0.67
Women	0.50	0.25	0.33

Mass. The mass of a row (row proportions) is the row total divided by the grand total. The mass of a column (column proportions) is the column total divided by the grand total. The average row profile is the *centroid* of the row profiles, where each profile is weighted by its mass. Column masses are derived similarly as shown in Exhibit 3.3. (The centroid or weighted average is known less commonly as the *barycentre*.) The notation generally used for row and column masses is shown in Exhibit 3.4.

EXHIBIT 3.3

Row (gender) and column (income) masses

	High income	Low income	Total	Row masses
Men	100	300	400	0.67
Women	100	100	200	0.33
Total	200	400	600	
Column masses	0.33	0.67		

EXHIBIT 3.4

Row and column masses

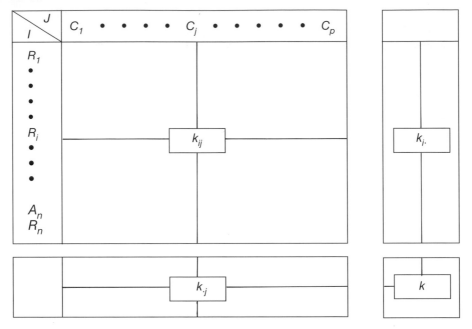

Inertia and display error. The accuracy of the correspondence analysis map is measured by 'percentage of inertia' (analogous to the expression 'percentage of variance explained' commonly used in multivariate analysis). The percentage of inertia not accounted for ('unexplained variance') in the map is considered as display error. The total inertia is the sum of the individual inertias of the rows when we are considering the set of row profiles, and as the sum of the individual inertias of the columns when we are considering the set of column profiles. Total inertia can also be split into components along the principal axes.

Contribution. Contribution refers to the inertia contributed by each row and each column to the total inertia. This tells the analyst the impact of different variables, thereby facilitating the interpretation of the axes.

The correspondence analysis model

In the correspondence analysis model, the initial data matrix, of the raw numbers of the contingency table, is transformed into two tables of profiles: a table of row profiles and a table of column profiles. With each of the profiles in each of these tables an appropriate mass is associated, and their centres of gravity are computed. To measure the distances between the profiles, a modified Euclidean distance called the distributional distance (or chi-squared distance) is used. Distributional distance is similar to the usual Euclidean distance except that, in computing chi-squared distances, we divide each squared difference between coordinates by the corresponding element of the average profile.

The dispersion of the profiles with respect to their centre of gravity is measured by inertia. Principal components analysis is applied to each of these tables to obtain the principal axes. The axes of one analysis are superimposed on the corresponding axes of the other analyses, two by two, to give perceptual maps in which both the variables of the data are plotted. These maps are interpreted using inertia components accounted for by the principal axes, and absolute and relative contributions (explained below) are obtained that are indispensable for understanding the structure of the data.

Although the central operation of a correspondence analysis is the determination of principal axes, correspondence analysis is not a simple variant of principal components analysis. Correspondence analysis can be applied to many other types of data besides contingency tables. One of the most useful extensions of correspondence analysis is to questionnaire data that are not in the form of a contingency table, but can be usefully analysed after an appropriate coding. The analysis of questionnaire data where each individual is described by his answers to many questions (variables) is known as multiple correspondence analysis.

Perceptual maps

The usual perceptual map of a correspondence analysis shows the axes two by two. For example, if only the first two axes are considered for interpretation, we can have a perceptual map showing axis 1 as the horizontal axis and axis 2 as the vertical axis. If three axes are used, then we can have a 1×2 map, a 1×3 map, and a 2×3 map, and so on. The principal maps of both the variables are plotted in this map.

The *quality* of representation of a row or a column on a principal axis is measured by its relative contribution (COR) in respect of that axis. If COR = 1000 (which is really equal to 1 since per mille values are printed in computer output) the quality of representation is perfect for that row or column on that axis. If COR = 0, then the row or column is not at all represented on that axis. Any intermediate value for COR shows the degree to which the row or column is present on that axis. (Some programs report relative contributions as proportions or percentages rather than as per mille values.)

In perceptual maps which take into consideration a certain number of axes, say two or three, the quality of representation is given by adding the COR values for these axes. If only two axes are considered, then Qual = COR1 + COR2; if three are considered, as in our hospital data later in this exposition, Qual = COR1 + COR2 + COR3.

Active and supplementary points

Sometimes, after an analysis has been completed, we may come across new data that could have been included in the analysis if we had had them earlier. For instance, suppose two distinct new brands of automobile are introduced into the market, after we have just completed an analysis showing the perceptual space of the consumers with respect to brands. It is then not necessary to redo the entire analysis in order to see how the new brands fit into the perceptual map already obtained. The transition formula is used to calculate the coordinates of these new brands and they are plotted on the same perceptual map. Their position in this map is an indicator of their relative place *vis-à-vis* the other brands. However, the COR values for these points are likely to be poor, and CTR (see Section 3.3) has no meaning for a supplementary element.

When a particular row or a column has a very large mass it is likely to overshadow the relationships that exist between the other rows or columns in the analysis. In such a case the analysis can be repeated by treating this element as supplementary. The analysis will then reveal more clearly the relationships between the other elements.

Multiple correspondence analysis

Typically, correspondence analysis deals with contingency tables, where data involve two qualitative variables. Questionnaires are, however, very common in market research, and typically they involve more than two, sometimes even several hundred variables. And these data do not constitute contingency tables. However, one of the most useful extensions of correspondence analysis is for the analysis of questionnaire data (see Exhibit 3.5). After appropriate recoding of these data, correspondence analysis can be applied to them. This is a very rich feature of correspondence analysis.

EXHIBIT 3.5

Multiple correspondence analysis input table

	AGE					HHLD INCOME			LANGUAGE	
	Under 25	25–34 yrs	35–49 yrs	50–64 yrs	65 or over	Under $30K	$30–$59K	$60K+	English	French
AGE										
Under 25	161	0	0	0	0	52	55	47	126	35
25–34 yrs	0	286	0	0	0	62	97	115	222	63
35–49 yrs	0	0	443	0	0	71	141	204	344	99
50–64 yrs	0	0	0	281	0	52	85	118	215	66
65 or over	0	0	0	0	228	82	66	47	180	49
HHLD INCOME										
Under $30K	52	63	70	52	82	139	0	0	224	96
$30K–$59K	55	97	141	85	65	0	442	0	340	102
$60K+	48	116	200	114	46	0	0	519	432	87
LANGUAGE										
English	105	185	266	170	132	180	273	342	858	0
French	42	77	117	76	57	108	119	115	0	368

The self-association of each variable (similar to a correlation of 1.0 on the diagonals of a correlation matrix), makes multiple correspondence analysis computationally efficient. However, the self-associations also increase the total inertia of the points besides complicating the usual geometric interpretation (Greenacre, 1993).

3.3 Correspondence analysis: computer output

In a study of online habits, 610 consumers were asked to indicate how frequently they use the Internet. The responses were then classified as low, medium, high, and very high. How is Internet usage related to different age groups? To answer this question, the data were tabulated against the age categories of the respondents and were subjected to correspondence analysis using the SimCA program. The output is given in Exhibit 3.6.

EXHIBIT 3.6

Internet usage and age output, annotated

DATA MATRIX *(our input matrix)*

	1 Vhigh	2 High	3 Med	4 Low	sum
1 <25y	99	41	11	5	156
2 25–34	73	97	49	17	236
3 35–44	39	25	29	9	102
4 45–54	15	9	11	9	44
5 55+	17	11	15	29	72
sum	243	183	115	69	610

ROW PROFILES *(Row proportions)*

	1 Vhigh	2 High	3 Med	4 Low	sum
1 <25y	63.5	26.3	7.1	3.2	100.0
2 25–34	30.9	41.1	20.0	7.2	100.0
3 35–44	38.2	24.5	28.4	8.8	100.0
4 45–54	34.1	20.5	25.0	20.5	100.0
5 55+	23.6	15.3	20.8	40.3	100.0
average	39.8	30.0	18.9	11.3	100.0

COLUMN PROFILES ROW PROFILES *(Column proportions)*

	1 Vhigh	2 High	3 Med	4 Low	average
1 <25y	40.7	22.4	9.6	7.2	25.6
2 25–34	30.0	53.0	42.6	24.6	38.7
3 35–44	16.0	13.7	25.2	13.0	16.7
4 45–54	6.2	4.9	9.6	13.0	7.2
5 55+	7.0	6.0	13.0	42.0	11.8
sum	100.0	100.0	100.0	100.0	100.0

INERTIAS AND PERCENTAGES OF INERTIA *(Variance explained by different axes or dimensions)*

Factor

	Eigenvalue	*Variance explained*
1	0.151448	67.19% ***
2	0.059976	26.61% *******************
3	0.013982	6.20% *****
	0.225407	*(Total)*

INERTIA CONTRIBUTIONS OF EACH CELL

	1 Vhigh	2 High	3 Med	4 Low
1 <25y	159	5	84	66
2 25-34	34	71	3	26
3 35-44	0	7	36	4
4 45-54	3	10	6	24
5 55+	35	38	1	388

Chi-square statistic (if applicable) = 137.50 (d.f.= 12)

(*The inertias reported here are on per mille scale i.e., we should be divide the numbers by 10 to convert them into percentages. For instance 15.9% of the inertia (159/1000) is accounted for by the first cell in the above table.) The chi-squared statistic of 137.50 for 12 degrees of freedom is highly significant.*

ROW CONTRIBUTIONS (*All numbers are on per mille scale*)

I	NAME	QLT	MAS	INR	k=1	COR	CTR	k=2	COR	CT
1	<25y	997	256	314	417	629	294	1319	368	434
2	25-3	920	387	134	74	69	14	257	851	427
3	35-4	127	167	48	124	9	1	87	118	21
4	45-5	895	72	42	1339	869	55	159	26	4
5	55+	992	118	462	1904	926	637	1240	65	113

COLUMN CONTRIBUTIONS

J	NAME	QLT	MAS	INR	k=1	COR	CTR	k=2	COR	CTR
1	Vhig	993	398	231	267	547	188	1241	445	386
2	High	829	300	131	149	227	44	243	602	296
3	Med	746	189	131	1219	306	60	262	440	216
4	Low	991	113	508	1974	937	708	1232	53	102

QLT (Quality): Proportion of information represented

MAS (Mass): Proportion of cases in corresponding rows/columns

INR (Inertia): Variance explained: obtained by adding the corresponding rows or columns in the previous table "Inertia contribution of each cell"

K = 1: Projection on the first factor (x-axis of the map)

K = 2: Projection on the second factor (y-axis of the map)

COR: The relative intertia accounted for (the contribution of the Kth axis to the inertia of the ith row)

CTR: The contribution of a cell to an axis. (For instance, for the first factor under 25 and 55+ contribute the most)

EXHIBIT 3.7

Correspondence analysis map of Internet usage

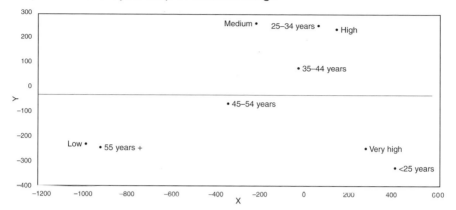

Interpreting the output

Based on the coordinates provided (under $k = 1$ and $k = 2$) we can derive a perceptual map (see Exhibit 3.7). Since these two dimensions together explain about 94% of the variance (see the output under 'Inertias and percentages of inertia'), we can conclude that a two-dimensional map will be a faithful representation of the contingency table.

To interpret the meaning of the axes, we need to interpret each axis separately. To interpret the first axis, let us drop all points directly to the x-axis (Exhibit 3.8). The points that are furthest apart in this dimension are the under-25 and 55+ age groups, implying that the dimension is defined by contrasting the youngest and oldest age groups. This can also be confirmed by inspecting the CTR under $k = 1$ in the computer output. In fact, as we move from left to right we see the groups becoming younger and younger. Turning our attention to heaviness of usage (the column variables), we note that once again the points are monotonic: the usage goes from low to very high. Obviously, there is a relationship between usage and age – heaviness of use goes with younger age groups. High and very high usage are on the right-hand side of the map. Clearly, the under-25 age group includes the heaviest users of the Internet. This can be confirmed by consulting the row profiles of the output.

To interpret the second dimension we do a similar analysis, but this time dropping all points on to the y-axis (Exhibit 3.9). Here the contrast is between under-25s and 25–34-year-olds when we consider age groups, and between 'very high' on the one hand and 'high' and 'medium' on the other when we consider heaviness of use. We cannot take into account the category 'low' because its COR for this axis is 0.053, which means it practically does not figure on this axis. For the same reason, namely poor values for COR, we cannot consider the age groups 35–44, 45–54 and 55+ on this axis.

Hence the interpretation of the perceptual map as a whole is that very high usage contrasts with very low on the one hand (first axis) and with high and medium on the other (second axis). Similarly, the under-25 age group contrasts with the over-55 group on the first axis and with middle age groups on the second axis. Extreme usages are associated with either ends of the age spectrum, while moderate usage levels are associated with middle age groups.

EXHIBIT 3.8

Interpreting the *x*-axis (first dimension)

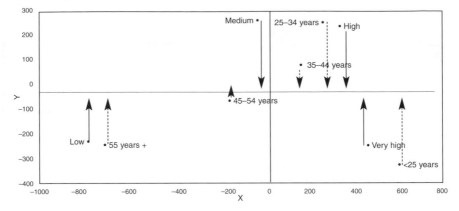

EXHIBIT 3.9

Interpreting the *y*-axis (second dimension)

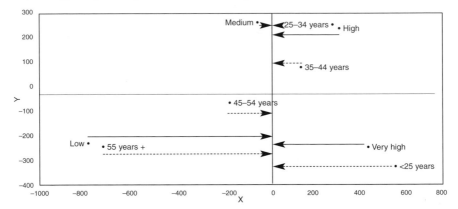

Because the data set used here is very small, the interpretation is fairly straightforward. The next section has examples of more complex data sets.

3.4 Correspondence analysis: marketing applications

Correspondence analysis has been finding increased use in marketing. It is being used in a variety of contexts as the following summaries of published studies show.

How to detect changes in behaviour using consumer panels

Marketing problem
Panels provide rich information to the marketer. Panels may consist of individual consumers, households, business customers, opinion leaders, or experts. The main advantage of panels is that they enable us to track changes over a period of time. Because panels consist of essentially the same respondents over a period of time, sampling differences are eliminated as a source of difference between time periods.

Yet, repeated measurements seldom show dramatic changes from one time period to another. They tend to be stable, especially over the short term. There may be minor changes, but they typically fall short of statistical significance.

One way to obtain meaningful information from the panel data is to *de-emphasize how much* change has occurred over the intervening period and *emphasize how* a given response changes over time. This way one can identify changes in trend over a period of time as opposed to computing the statistical significance of minor observed differences.

In this example, Thiessen, Rohlinger and Blasius (1994) demonstrate how minor changes in panel data may be interpreted by using the data collected by Eckart and Hahn concerning married women's integration into the labour market. In the 1950s and the 1960s only a minority of women were in the workforce. Men were mainly responsible for external (to the household) activities and women were responsible for internal activities. Subsequently, the participation of women in the labour force began to accelerate. How did this change affect the division of household labour? To answer this question, the investigators chose to apply correspondence analysis to panel data.

Application of correspondence analysis

In 1977, 1978 and 1980, 223 two-income couples (newly-wed and childless when recruited) from five cities in Germany were asked who was responsible for the following tasks:

1. Preparing breakfast (B)
2. Making dinner (D)
3. Preparing main meals (M)
4. Minor household repairs (R)
5. Shopping (S)
6. Laundry (L)
7. Insurance matters (I)
8. Automobile driving (A)
9. Washing dishes (W)
10. Financial matters (F)
11. Tidying the house (T)
12. Official matters such as taxes (O)
13. Vacation planning (V)

Although information on these 13 tasks was collected from husband and wife independently, only those responses in which there was agreement between the two were included in the analysis.

The strategy adopted by Thiessen *et al.* was to use the 1977 data to identify the primary dimensions on the map and plot the subsequent data (1978 and 1980) on the map so created. To achieve this, they treated the 1978 and 1980 data as supplementary points in the analysis. This created five conceptual 'sections' in the analysis, as shown in Exhibit 3.10. (Although the analysis started with dual-income families, the design allowed for change in employment status in subsequent time periods. This enabled the researchers to study the impact of change in employment status on gender roles.) Each section contains parallel information on the respondents: each row represents a household task and the numbers of respondents who thought that that particular task was the wife's responsibility (WW), the husband's responsibility (HH), joint responsibility (JJ) or alternating responsibility (AA).

EXHIBIT 3.10

Analysis design

Wave 1 Section 1		Reference frame
Wave 2 Section 2a (Wife in workforce)	Wave 2 Section 2b (Wife not in workforce)	Supplementary rows
Section 3a (Wife in workforce)	Section 3b (Wife not in workforce)	Supplementary rows

Results of correspondence analysis

Thiessen *et al.* applied correspondence analysis to the household data and derived a two-dimensional symmetric map (Exhibit 3.11).[1]

EXHIBIT 3.11

Correspondence map of household panel data

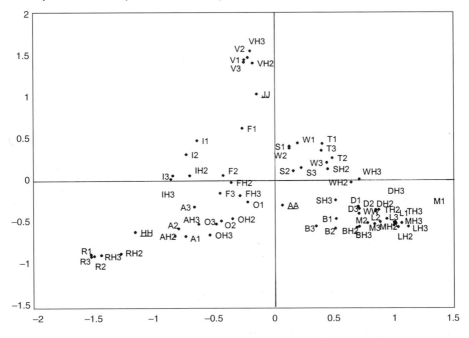

---

[1]The correspondence maps in Exhibits 3.11 and 3.12 were created by me based on an independent analysis of the data in the original paper. Therefore they may not exactly correspond to the maps in the original paper. However, the differences are generally minor and have no implications for interpretation.

Interpreting the axes. Looking at Exhibit 3.11, we note that on the left we have HH and on the right WW. This suggests that the first axis (horizontal) has to do with agreement on gender roles – which tasks are done by the wife and which ones by the husband. Using similar logic we find that the second axis (vertical) distinguishes alternating tasks (AA) from tasks that carry joint responsibility (JJ). The first axis explains 49% of the total variance, while the second axis explains 40%. Since both axes taken together explains nearly 90% of the total variance, the two-dimensional map is taken to be an excellent representation of the data.

Also, in this analysis the first and second axes seem to be equally important since they account for nearly equal amounts of the inertia. It is then not necessary to interpret the points separately on each axis.

Interpreting the data points. To understand how the responsibility for a given task changes over time, we can connect the three time periods for any given task, as is done in Exhibit 3.12. Let us consider two examples, shopping and insurance. Shopping (S), initially the wife's responsibility (as it is on the right-hand side of the map), moves further to the right and down (towards WW and AA). It then moves more to the right and somewhat upwards, thus consolidating the initial gender role as time goes by, and moving towards alternate responsibility from wife's sole responsibility. Similarly, insurance (I), initially the husband's responsibility (located on the left of the map), moves steadily down and to the right, further consolidating the initial gender role, but showing a tendency to AA.

It is also clear that the gender roles of households where the wife no longer works (tasks ending with H2 or H3) become consolidated. The wife becomes more responsible for home-related tasks, as evidenced by the bunching at the right-hand end of the map of H2 and H3 tasks that are internal to the household. In fact, the direction of the lines for household tasks tends to move downward to the right, indicating that what was joint responsibility when the wife was working becomes her sole responsibility when her employment status changes.

How correspondence analysis addressed the problem

When we are dealing with consumer panel data, the changes we observe are gradual. Most products and brands are in a 'steady state', meaning that unless something special happens, data collected at one time period tend to look fairly similar to data collected at another time period. Traditional methods of analysis such as calculating statistical significance of differences between two time periods are likely to be unproductive.

By using correspondence analysis with supplementary points, Thiessen *et al.* managed to identify recurring patterns, even though these patterns in and of themselves are minor and likely not statistically significant. Their use of correspondence analysis enabled them to identify many patterns, among them the following

1. There is a strong gender bias in task assignment within a household;
2. These biases become consolidated when the wife ceases to work;
3. Wife's employment status has greater impact on wife's tasks than on husband's task.

Although the above study is more concerned with the gender roles and thus has indirect effect on marketing issues, the methodology of Thiessen *et al.* can be applied to regular consumer panels that deal with purchase behaviour.

EXHIBIT 3.12

Correspondence map of household panel data, with time period connections

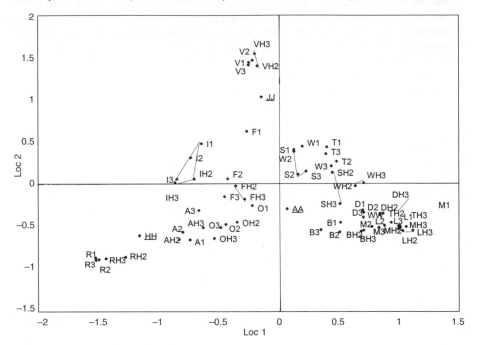

How do consumers decide what to buy?

Marketing problem

How do consumers decide which product to buy? Standard economic theories assume that consumers try to maximize their utility. To do so, consumers must have access to relevant information – real or perceived – which will then enable to them to evaluate alternative products. Consumer evaluations of products are assumed to be based on an overall comparison of alternative products as well as an evaluation of the relative importance of product attributes.

In many product categories, such as desktop computers and telephones, technical specifications converge – any manufacturer can produce a computer of a given specification and price. Even though it is less obvious, this is true of many consumer goods. In such cases, the non-technical aspects of the product such as product design and product image may assume a crucial role in persuading the consumer to choose one brand over another.

One major component of non-technical specification of a product is its overall design. Telephones, which perform identical functions equally well, can come in different colours and shapes. Marketers may be interested in understanding how consumers group the products on the basis of overall design. Another related question is whether consumer judgements in this context may be used to identify consumer preferences. However, instead of conjuring up the basis on which consumers group the products and speculating whether the groups relate to their (consumers') preferences, we can ask the consumers to state how they would name the groups of products they created. Once we have such data, we can analyse the data using correspondence analysis, as Snelders and Stokmans (1994) illustrate below.

Application of correspondence analysis

To understand how consumers group telephones of different design, Snelders and Stokmans sought the views of 43 members of a consumer panel of the Delft University of Technology (Netherlands). The panel was a random sample of the residents of Delft and surrounding areas. Seven telephones (the then largest-selling models) were used in the study. Four of these were chosen for their diversity of appearance and the remaining three were chosen randomly.

The chosen consumers were asked to make distinctions among sets of telephones on the basis of their design, assuming that the price and quality of all products were identical. They were shown all seven models and asked to 'split the group of telephones and denominate each group'. In all, the consumers generated 516 denominated groups of telephones. For purposes of analysis, these 516 groups were categorized into 50 distinct attribute categories. This resulted in a contingency table (7 brands and 50 attributes) in which each cell represented the number of times an attribute was applied to a given brand. This table was subjected to correspondence analysis and symmetric maps were derived.

Results of correspondence analysis

A three-dimensional solution, which explained 74% of the variance, was chosen for further interpretation. These maps are shown in Exhibits 3.13 and 3.14.[2]

Axis 1 (business vs. residential). On the right-hand side we have telephones T1, T4 and T5, and on the left, telephones T2, T6 and T7. T3 occupies a neutral position, right in the middle. To understand why the telephones are positioned this way, we look at the attributes at the extreme ends of this dimension. On the right, we find attributes such as 'living room', 'colourful', 'small', 'eye-catching', and 'elegant', while on the left we find attributes such as 'big', 'businesslike', 'complicated', and 'reliable'. So the first major distinction made by consumers on the basis of appearance is between large business phones and small residential phones.

Axis 2 (vertical vs. horizontal receiver). Models T1, T2, T4 and T6 are in the top half of the map and Models T3 and T7 in the bottom half, with T5 occupying a neutral position on this axis. Statistical tables (inertia) showed that models T2 and T6 versus T3 and T7 have been most influential in deciding this axis. In the top half we have attributes such as 'vertical receiver', 'complicated' and 'eye-catching' and at the bottom half attributes such as 'horizontal receiver', 'uncomplicated' and 'sombre'. While it was somewhat difficult to interpret this axis, the investigators concluded that that the consumers' judgements were swayed mainly by the placement of the receiver: vertical vs. horizontal.

Axis 3 (avant-garde vs. classical). The major contrast in this dimension (from the inertia table) was between models T1 and T4. The contrast was between attributes like 'oblique angles', 'small', and 'modern' versus attributes like 'classic', 'ugly' and 'complicated'.

---

[2]As with Exhibits 3.11 and 3.12, these correspondence maps were created by me based on an independent analysis of the data in the original paper. While they may not exactly correspond to the maps in the original paper, the differences are generally minor and have no implications for interpretation.

EXHIBIT 3.13

Correspondence analysis map (axis 1 vs. axis 2)

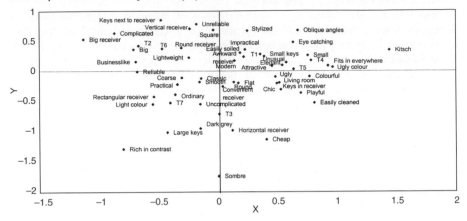

EXHIBIT 3.14

Correspondence analysis map (axis 1 vs. axis 3)

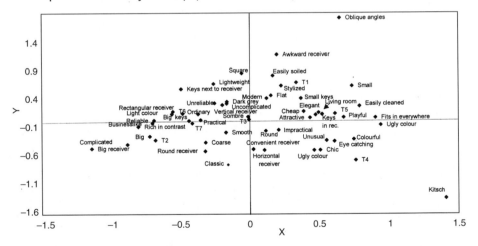

How correspondence analysis addressed the problem

By using correspondence analysis, Snelders and Stokmans effectively answered many questions.

1. It is possible to apply correspondence analysis to marketing problems where no set questionnaire with predetermined attributes is used. (Although maps can be derived from the way consumers group products using techniques such as non-metric multidimensional scaling, the investigators' methodology enables us to identify the underlying dimensions of distinction much more readily.)

2. The three dimensions identified by the analysis are logical and provide insight into the way brand (model) imagery works. The interesting aspect of these

distinctions is that, while consumers classify the models based on obvious and dominant features such 'business vs. residential', they then proceed to add image characteristics ('eye-catching', 'complicated', 'cheap', etc.) to these dimensions.
3. In the first two dimensions 'ugly' and 'attractive' are positioned closely together. In the third dimension, 'ugly' contrasts with 'modern' rather than with 'attractive'. This addressed an important issue. These maps reflect product positioning and not product preference.

This example illustrates that correspondence analysis can be used to analyse non-standard data, and perceptual maps cannot automatically be assumed to tell us anything about consumer preferences.

How to assess corporate image

Marketing problem

Corporations spend considerable sums of money to create a positive image of themselves. A positive corporate image is one of the major assets of a corporation since it is assumed that a positive corporate image can be transferred to the products and services of the corporation (which in turn might reinforce the positive corporate image).

Corporate image does not exist in a vacuum. Since businesses in capitalist economies by their very nature are competitive, it is not enough to measure corporate image in isolation. For instance, Corporation X may be considered a good corporate citizen, but other competing corporations may be considered even better in this respect. Consequently, to understand where a particular corporation fits in, we need to measure the images of competing corporations as well.

When we collect data on competing corporations, it would be helpful to have the information summarized in the form of a perceptual map, so the image of a particular corporation can be visually related to those of others. In the example that follows the Lake Hospital System (LHS) wanted to understand its corporate image in relation to its competitors. Javalgi *et al.* (1992) used correspondence analysis for this problem.

Application of correspondence analysis

Javalgi *et al.* started with the premise that healthcare consumers draw conclusions about an institution's overall image from impressions they form about the strengths and weaknesses of the hospital's offerings. Consumers form images from their past experiences, word of mouth, and marketing communications. Understanding the features that consumers associate with a hospital can be used as an effective image management tool. Such image management can be used to ensure public support, attracting favourable legislation, gaining tax breaks, and outside funding. Image management can be crucial for strategic planning.

A well-designed study can help the marketer to differentiate the images of different hospitals. However, consumers cannot really be expected to rate or rank all hospitals or hospital attributes. Neither can they be expected to judge directly how different hospitals are similar or different. Such tasks are either too cumbersome or too confusing for the respondents.

Consequently, Javalgi *et al.* chose correspondence analysis to simplify the data collection process and provide a joint graphic display of the relationship between different hospitals and image characteristics associated with them.

For the study itself, they first developed a set of 13 attributes to represent hospital images (see Exhibit 3.15), based on focus group interviews and literature review. On this basis, they constructed a formal questionnaire and contacted by telephone 503 geographically dispersed respondents in the market area using random probability sampling procedures. Respondents were then told that a number of hospital features would be read out to them and they were to indicate which hospitals are associated with each characteristic.

The above procedure provided the data in the form of contingency table: how many people thought a given hospital possessed each of the 13 characteristics. This served as the input for correspondence analysis.

EXHIBIT 3.15

Perceived features of hospitals

HOSPITALS

Features	CC	GC	GV	LHS	LE	LW	MER	EU	HC	HR	MH	MS	RH	SL	SV	UH
emer	76	17	12	35	94	108	14	49	39	17	43	19	23	13	18	58
hart	260	5	3	15	37	44	8	25	17	22	11	13	6	5	12	66
rehb	156	9	7	25	64	66	17	69	18	12	19	12	14	9	12	44
canc	201	5	3	9	36	29	5	26	14	10	7	12	6	9	4	104
call	44	6	4	24	68	57	13	30	17	5	6	9	9	6	6	31
womn	60	17	15	25	121	60	15	76	47	17	14	18	15	12	12	74
lasr	112	7	6	17	35	27	33	63	36	12	9	21	10	10	8	48
outp	81	29	24	53	146	131	35	91	70	32	27	30	31	26	27	67
docs	220	37	30	36	86	87	35	70	48	37	39	47	47	32	34	123
attn	90	32	26	35	120	101	24	74	56	34	27	32	38	26	26	69
snrs	32	11	8	32	73	82	16	44	18	12	9	11	27	8	8	24
comm	54	15	11	60	126	127	24	47	30	15	13	13	23	12	12	35
tech	298	13	10	17	44	57	16	38	26	14	17	34	19	14	12	105

Legend

Features

emer	expert emergency treatment
hart	heart disease prevention and treatment
rehb	rehabilitation services
canc	cancer treatment
call	call-in health information services
womn	women's health services
lasr	laser surgery
outp	outpatient services
docs	doctors keep up with medical advances
attn	staff giving personal attention
snrs	special programs for seniors
comm	Offering community programs
tech	advanced technological equipment

Hospitals

CC	Cleveland Clinic
GC	Geauga County Hospital
GV	Geneva Hospital
LHS	Lake Hospital System
LE	Lake East Hospital
LW	Lake West Hospital
MER	Meridia Hospital System
EU	Meridia Euclid Hospital
HC	Meridia Hillcrest Hospital
HR	Meridia Huron Road Hospital
MH	MetroHealth Medical Centre
MS	Mt. Sinai Hospital
RH	Richmond Heights
SL	St. Luke's Hospital
SV	St. Vincent Charity Hospital
UN	University Hospital

Results of correspondence analysis

Exhibit 3.16 shows the absolute contributions to variance by different attributes and different hospitals. This is simply the percentage of variance explained by each item in relation to each axis (principal components). The larger the absolute contribution of an item to an axis, the more important is that item in determining the axis. For the first axis, we note that items 'heart disease prevention and treatment' (21.4%), 'cancer treatment' (18.8%), and 'advanced technological development' (19.7%) together represent 60% of the explained variance of the first dimension. We can thus consider the first dimension to be 'specialized treatment'. For the second axis, the two attributes 'doctors who keep up with medical advances' (25%) and 'offering community programs such as support groups, classes and health screenings' (26.8%) account for 51% of the variance. We can thus consider the second dimension to be 'keeping up'.

EXHIBIT 3.16

Absolute contributions to variance by rows and columns

Features (Row Variables)	Factor 1	Factor 2
Expert emergency treatment	3.2	0.3
Heart disease prevention and treatment	21.4	7.5
Rehabilitation services	0.9	3.8
Cancer treatment	18.8	1.3
Call-in health information services	1.9	8.9
Women's health services	3.1	1.7
Laser surgery	0.8	7.4
Outpatient services	8.6	1.7
Doctors keep up with medical advances	0.7	5.0
Staff giving personal attention	4.3	1.2
Special programs for seniors	6.6	5.3
Offering community programs	9.9	5.8
Advanced technological equipment	19.7	0.0

Hospitals (Column Variables)	Factor 1	Factor 2
Cleveland Clinic	60.4	5.1
Geauga County Hospital	1.1	6.6
Geneva Hospital	1.0	8.0
Lake Hospital System	4.4	11.0
Lake East Hospital	11.5	9.0
Lake West Hospital	8.1	22.7
Meridia Hospital System	0.7	1.7
Meridia Euclid Hospital	1.9	1.7
Meridia Hillcrest Hospital	1.9	7.2
Meridia Huron Road Hospital	0.0	8.0
MetroHealth Medical Center	0.3	2.4
Mt. Sinai Hospital	0.1	2.4
Richmond Heights	1.7	1.9
St. Luke's Hospital	0.3	5.6
St. Vincent Charity Hospital	0.3	3.7
University Hospital	6.2	2.1

The perceptual map produced by correspondence analysis is shown in Exhibit 3.17. The horizontal axis explains 71% of the total variance while the vertical axis explains 11%, for a total of 82%, which can be considered a reasonable representation of the original data represented by the contingency table. Many inferences can be drawn from this map:

- Cleveland Clinic is closely associated with 'heart disease prevention and treatment', 'cancer treatment', and 'advanced technological development'.
- University hospitals also project a similar image, but to a much lower extent.
- A large number of hospitals – Euclid, Meridia, Metro Health, Richmond, Huron, Hillcrest, St. Vincent, St. Luke's, Geauga, Geneva – fall in the top left-hand quadrant of the map, which indicates that they are associated with 'keeping up' but not with 'specialized treatment'.
  √ Meridia's image does not seem to be distinct from the images of Metro Health, Richmond Heights and St. Vincent Charity.
  √ Euclid has a more distinct image, it being associated with 'outpatient services' and 'women's health services'.
  √ All hospitals in this quadrant are also perceived to have 'staff giving you personal attention'.
  √ Although 'laser surgery' is a common theme of Meridia advertising, the map shows that Meridia is not differentiated on this dimension.
  √ Mt. Sinai is differentiated from Meridia in being associated with 'doctors who keep up with medical advances'.

EXHIBIT 3.17

Correspondence analysis of hospital features

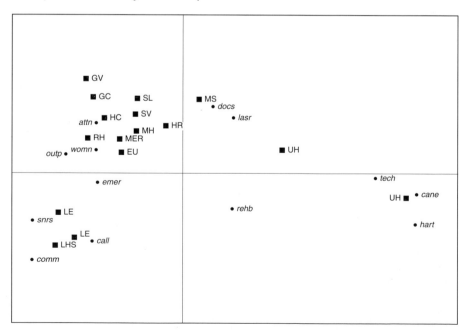

- Lake West and Lake East are perceived to be a cohesive healthcare system with a distinctive image. While both of them are perceived to be community-oriented, Lake East is distinguished by being associated with 'programs for seniors', while Lake West in distinguished by being associated with 'call-in information'.

How correspondence analysis addressed the problem
The use of correspondence analysis helped the investigators accomplish both their major marketing objectives: eliciting valuable corporate image evaluations without making the task onerous to the respondents, and obtaining the image characteristics of different hospitals – their similarities and differences.

Since the input required for correspondence analysis is much less complex than rating each of the 16 hospitals on each of the 13 features (a task involving $13 \times 16 = 208$ separate judgements), the data were collected with relative ease. The task simply involved a 'yes' or 'no' response for each feature in relation to the 16 hospitals, thus grossly simplifying the respondent's task.

With regard to the second objective – eliciting relevant corporate image information – correspondence analysis provided rich information. For instance, it showed that the hospitals clustered into three image groups (treatment-oriented, patient-oriented, and community-oriented); that differences in image exist among the hospitals within each of the three; and that Cleveland Clinic has a very distinct image. In addition, correspondence analysis also provided an easy-to-communicate visual representation of the corporate images of different hospitals.

Even more importantly, the results of this study helped the sponsor – LHS could use the results for strategic purposes. In areas where LHS had strong internal clinical strength, but this was not perceived by consumers, suitable communications programs were developed; in areas where LHS had strong internal clinical strength, and this was perceived by consumers, new markets were created; and in areas where LHS was clinically weak, defensive measures were undertaken by comparing its strengths with competitors' weaknesses. Such strategies helped LHS to define its image clearly and position itself in the marketplace more distinctively.

3.5 Caveats and concluding comments

Perceptual maps collapse multidimensional data into fewer dimensions, usually two. This necessarily results in some loss of information. The trade-off between information loss and ease of interpretation is common to all data reduction techniques. However, perceptual maps are particularly vulnerable to erroneous interpretation since the interpretation of a map appears 'intuitive' and straightforward, even when it may not be. Hence it is absolutely essential to rely on the printed values of contributions, relative (COR) and absolute (CTR), in interpreting a perceptual map.

'Over-interpreting' the positions

When interpreting the map, it is important to remember that the map provides the best possible representation, given the nature of the underlying data. A two-dimensional map that explains 88% of the variance is a better representation of the data than a map that explains only 73%. Since there is no cut-off point that is used as a criterion for accepting a map, the stability associated with points on the map (e.g., where a brand is positioned) may vary from one map to another. Moreover, the map is wholly dependent on the variables and brands included. The addition or deletion of a brand or attribute can influence the positions. Consequently, it is best to treat the attribute and brand positions on a perceptual map as a good summary of relative positions rather than a precise reflection of the market.

Directly relating row and column variables

Correspondence analysis, as noted earlier, produces two different maps – one for the row variables one for the column variables – and overlays one map on another. While a precise interpretation is possible for each set (e.g., set of brands or set of attributes), the two sets can only be related directionally. Thus to interpret the closeness between an attribute and a brand could be tricky, and lead to serious error, except when both an attribute and a brand constitute the extreme points on an axis. Although a suggestion to normalize the map so the two sets can be compared has been proposed and implemented by Carroll, Green and Schaffer (1986, 1987, 1989), some experts in this area such as Greenacre (1989, 1993) consider it an incorrect procedure.

In the English-speaking world, correspondence analysis is used essentially as a tool for creating perceptual maps, while in France the technique is used as a tool for more complete data analysis. French statisticians such as Benzécri, Lebart, Jambu, and Volle use correspondence analysis (with cluster analysis) as a primary tool for interpreting data. We have followed here the tradition of viewing the technique as a tool for creating perceptual maps. Those who are interested in using correspondence analysis as a more serious data analytic tool will benefit from reading Benzécri (1992), Lebart, Morineau and Warwick (1984) and Jambu (1991).

Bibliography

Further reading

For a thorough exposition of correspondence analysis and its analytic capabilities readers are referred to Benzécri (1992), Jambu (1991) and Lebart *et al.* (1984). An excellent but much less technical introduction can be found in Greenacre (1993).

References

Benzécri, Jean-Paul (1992) *Correspondence Analysis Handbook.* Marcel Dekker, New York.

Carroll, J.D., Green, Paul E. and Schaffer, C.M. (1986) Interpoint distance comparisons in correspondence analysis. *Journal of Marketing Research*, **23**, 271–80.

Carroll, J.D., Green, Paul E. and Schaffer, C.M. (1987) Comparing interpoint distance comparisons in correspondence analysis: A clarification. *Journal of Marketing Research*, **24**, 445–50.

Carroll, J.D., Green, Paul E. and Schaffer, C.M. (1989) Reply to Greenacre's commentary on the Carroll–Green–Schaffer scaling of two-way correspondence analysis solution. *Journal of Marketing Research*, **26**, 366–8.

Greenacre, Michael J. (1989) The Carroll–Green–Schaffer scaling in correspondence analysis solution: A theoretical and empirical appraisal. *Journal of Marketing Research*, **26**, 358–65.

Greenacre, Michael J. (1993) *Correspondence Analysis in Practice.* Academic Press, San Diego, CA.

Jambu, Michel (1991) *Exploratory and Multivariate Data Analysis.* Academic Press, San Diego, CA.

Javalgi, Rajshekhar, Whipple, Thomas, McManamon, Mary and Edick, Vicki (1992)

Hospital image: A correspondence analysis approach. *Journal of Healthcare Marketing* (December), 34–8.

Lebart, L., Morineau, A. and Warwick, K.M. (1984) *Multivariate Descriptive Statistical Analysis: Correspondence Analysis and Related Techniques for Large Matrices*. Wiley, New York.

Snelders, H.M.J.J. and Stokmans, Mia J.W. (1994) Product perception and preference in consumer decision-making. In M. Greenacre and J. Blasius (eds), *Correspondence Analysis in the Social Sciences*. Academic Press, San Diego, CA, pp 252–66.

Thiessen, V., Rohlinger, H. and Blasius, J. (1994) The 'significance' of minor changes in panel data: A correspondence analysis of the division of household tasks. In M. Greenacre and J. Blasius (eds), *Correspondence Analysis in the Social Sciences*. Academic Press, San Diego, CA, pp 324–49.

4

Cluster Analysis

4.1   What is cluster analysis?

Segmenting customers into homogeneous groups is one of the basic strategies of marketing. How do we group consumers who seek similar benefits from a product so we can communicate with them better? How do we group the financial characteristics of companies so we may be able to relate them to their stock market performance? Questions like these require that we group items on the basis of their similarities. Cluster analysis is a group of techniques frequently used for this purpose.

The aim of cluster analysis is to group objects that are similar on a set of predetermined characteristics. For instance, we may want to group consumers ('objects') based on their attitudes towards making expensive purchases. Or we may want to group different brands of laptop computers based on how they are perceived by consumers. (Since cluster analysis is used primarily for grouping respondents in marketing, I have often used the term 'individual' rather than the generic term 'object' in this chapter.)

There are two basic ways to accomplish this: *hierarchical* methods and *non-hierarchical* methods. In hierarchical methods, clustering at each level is dependent upon what happened at the previous level. For instance, if we have three groups and we want to create four groups, then one of the three groups is split into two; conversely, if we have four groups and want to make them into three, we do this by combining two of the four groups. In non-hierarchical clustering methods, clustering at a given level does not depend on what happened at the previous level. For example, a four-cluster solution is created directly from the data rather than from a three- or five-cluster solution. Non-hierarchical cluster solutions at different levels may not have any discernible relationship to clusters created at the previous level.

The purpose of cluster analysis is to form groups of objects such that, with respect to clustering variables, each group is as homogeneous as possible and as different from the other groups as possible. While this sounds like a fairly straightforward requirement, in reality there is no single ideal solution for creating clusters. Many decisions such as the measurement of similarity or dissimilarity, the type of clustering method (hierarchical or non-hierarchical), the actual clustering technique (there are many specific techniques within each type), and the number of groups selected are, in fact, heuristic in nature. Consequently, it is quite possible to come up with different sets of clusters depending on the heuristics used. It can be shown that one can create clusters even out of random numbers (which, by definition, cannot have any underlying groups). Consequently, the usefulness of clustering is seldom decided by its statistical correctness rather than by its practical utility.

Cluster analysis is used in marketing research primarily to segment the market. There are other, less widely used applications as well. Here are some marketing applications of cluster analysis:

1. A car manufacturer believes that buying a sports car is not solely based on one's means or on one's age. Rather, it is a lifestyle decision. Sports car buyers have a pattern of lifestyle that is different from those who do not buy sport cars. The manufacturer would like to identify the segment with a lifestyle that correlates most with buying sports cars to create a focused marketing campaign.
2. A women's clothing manufacturer believes that not all women buy clothes for the same reason. Some may buy clothes because of the price, some because they are in fashion, and yet others because they are durable. The manufacturer would like to identify the different market segments based on expected benefits (benefit segmentation). This can then be used to create a specific image for the company based on a particular benefit.
3. A marketing manager would like to group different brands of cereals that are perceived similarly by consumers so as to understand the segment in which her brand is competing.
4. An insurance company would like to segment its sales force based on factors that motivate them to sell. The company hopes to use this information to motivate the employees with different incentives that would appeal to different segments.

4.2 Cluster analysis model

To cluster a group of objects (e.g., consumers), we start with a set of predetermined clustering variables x_1, x_2, x_3, ..., x_n (e.g., consumers' lifestyle characteristics). Our objective is to divide the consumers into k homogenous groups such that each group is as dissimilar as possible from the remaining groups. All cluster analysis procedures involve:

1. a method for determining how similar two individuals are;
2. overall procedures for creating segments (hierarchical and non-hierarchical procedures); and
3. specific techniques for creating clusters.

Measuring similarity (distance measures)

Our first task is to determine how close consumer A is to consumer B, given their profiles on the clustering variables x_1, x_2, x_3, ..., x_n. All clustering programs use some kind of similarity (or distance) measure to decide how close two objects are. The most commonly used distance measure is the *Euclidean distance*, in which the difference between two consumers is calculated as the square root of the sum of squared differences of all clustering variables:

$$D_{ij} = \sqrt{\sum_{k=1}^{p} (X_{ik} - X_{jk})^2}$$

where D_{ij} is the distance between consumers i and j and p is the number of clustering variables.

If we simply add up the absolute difference between i and j on all p variables we obtain the *Manhattan distance,* also known as the *city block distance*:

$$D_{ij} = \sum_{k=1}^{p} |X_{ik} - X_{jk}|.$$

Both the Euclidean and Manhattan distance are special cases of the *Minkowski distance,* given by

$$D_{ij} = \sqrt[n]{\sum_{k=1}^{p} (X_{ik} - X_{jk})^n}.$$

For the Euclidean distance $n = 2$, and for the Manhattan $n = 1$.

Although the Euclidean distance is the most frequently used distance measure, there are two problems with it. First, it is scale-dependent. The distance between two individuals may vary, dependent on the units in which the clustering variables are expressed. For instance, if we cluster individuals on the basis of their height, weight, and income, whether the height is expressed in inches or in centimetres will decide how close two individuals are. This problem can be overcome by standardizing the data. However, there is a price to pay. Standardizing the data minimizes group differences since the large variances of distinguishing variables will be equalized when the variables are standardized. Second, the Euclidean distance does not account for correlations among the clustering variables. If some of the clustering variables are highly correlated, it amounts to 'double counting' when we compute the distance. The Mahalanobis distance D^2 is scale-invariant and takes into account the correlations among variables. There are some theoretical reasons why D^2 is not a suitable measure for computing distances in cluster analysis. Consequently, D^2 is not available in the clustering routines of most major statistical packages such SPSS and SAS (but see Section 6.2).

EXHIBIT 4.1

Distance measure for binary coefficients

	Respondent i	
	Yes	No
Respondent j Yes	a	b
No	c	d

Matching coefficient	$(a+d)/(a+b+c+d)$
Jaccard coefficient	$a/(a+b+c)$
Sørensen's per cent similarity	$200\, a/(2a+b+c)$
Sokal and Sneath	$a/[(a+2(b+c)]$

While correlations are a measure of similarity, they lack the properties of a true metric and are seldom used in cluster analysis. Special distance measures are available for binary data, as shown in Exhibit 4.1. For both metric and non-metric data there are many other (less frequently used) distance measures as well.

Forming clusters

Our second task relates to deciding on rules for forming clusters. How do we divide a group of individuals into *k* clusters? There are many ways assigning individuals to clusters.

Hierarchical clustering

Hierarchical clustering commonly uses one of four different procedures to create clusters.

Single-linkage clustering (nearest neighbour). In the following illustration, let us assume that we have two clusters. A, B, C belong to cluster I and D, E, and F belong to cluster II.

Now we need to assign the next individual G to one of the two clusters.

There are many ways in which we can decide whether G belongs to cluster I or cluster II. One way is to calculate the distance between G and the nearest point in cluster I (C) and the nearest point in cluster II (D) and assign the new individual to the closer cluster. In this case we will assign G to cluster II since G is closer to D than to C. This technique is called the nearest-neighbour method. Once we find the nearest neighbour of a new point, we are not concerned at all about the distance between the new point and all other points in a cluster. In our example, our only concern is to understand how far G is from D, not how far G is from E or from F. If a point is not the nearest neighbour, it plays no role in the formation of a cluster.

EXHIBIT 4.2
___

Euclidean distances between five individuals

	A	B	C	D
B	3.7			
C	1.4	3.5		
D	5.4	3.3	6.2	
E	3.0	4.6	7.8	7.1

Once we have a distance measure (Euclidean distance) and an assignment rule (nearest-neighbour method), we are ready to carry out cluster analysis. Exhibit 4.2 shows the Euclidean distance between five individuals.

First we look for the two respondents who are closest. We note that of all respondents, A and C are closest to each other – with a distance of 1.4, the smallest in the table. So let us combine them into cluster 1.

The next pair that are closest to each other are E and A (distance 3.0). (Note that A and C have already formed a cluster.) A is the nearest neighbour for the next point E. So we combine it with A and C.

We look for the next closest pair: it is B and D (distance 3.3). Since neither B nor D is already in a cluster, we create a new cluster that consists of B and D:

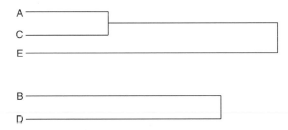

The next closest pair are B and C (distance 3.5). B and C are in different clusters, but they happen to be the nearest pairs, so we join both clusters.

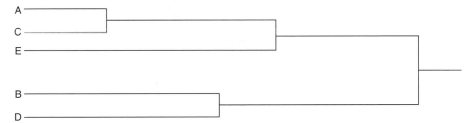

The above representation is called a *dendrogram*. From this you will note several things. First, we start with everyone being separate and gradually combine individuals into larger and larger and clusters until we end up with one single cluster. Secondly, the dendrogram preserves the original distances of the matrix as much as possible. Third, since each cluster is formed by annexing more individuals, we can have as many or as few clusters as desired. For instance, if we want to group the respondents into two groups, we can start from the left of the dendrogram, start moving right and stop at the point where there are two branches. In our case the clusters will be {A, C, D} and {B, E}. If we need three clusters we go further to the right until we encounter three branches: {A, C}, {D} and {B, E} and so on.

The nearest neighbour method is also known as *single-linkage clustering*. It is not problem-free. Let us return to our earlier example.

A B C	G D E F
Cluster I	*Cluster II*

Here we combined G with cluster 2 because it was close to D than to C. Now watch what happens when we have another new point H.

A B C	H G D E F
Cluster I	*Cluster II*

The nearest-neighbour logic would have H combine with G, since H is closer to G than it is to C. Now assume we have another point J, as shown below:

A B C	J H G D E F
Cluster I	*Cluster II*

The point J will be combined with cluster II because it is closer to H than it is C. By now the problem should be obvious.

A B C	J H G D E F
Cluster I	*Cluster II*

If you examine the original clusters {A, B, C} and {D, E, F}, clearly many points such as J and H are closer to A, B and C than to D, E and F. Again, if point G were not in the sample, J and H would be combined with cluster I rather than with cluster II. Yet, because the nearest-neighbour method concentrates on the closest point, we may end up with clusters that may not represent reality very well. This problem is known as the *linking* or *chaining effect*.

The farthest-neighbour method (Complete linkage) To solve the problem of the linking effect, we could measure closeness from the *farthest point* in each cluster rather than from the nearest point. If we apply this logic to the problem we will get the following configuration because we will be assigning new points depending *on how close it is to the farthest point* in each cluster (A for cluster I and F for cluster II).

A	B	C	J	H	G	D	E	F
	Cluster I				Cluster II			

While this seems to be an improvement over the nearest-neighbour method, it is not very satisfactory either. Assigning {A, B, C} to one cluster and {J, H, G, D, E, F} to another cluster might have been more logical, at least from a visual perspective, as shown below:

A	B	C	J	H	G	D	E	F
	Cluster I				Cluster II			

Since, as before, a single point (farthest point) decides how the groups are formed, a single point to the left of A could change the way the two groups are formed.

Centroid clustering To avoid depending on a single point, we can use *average-linkage cluster analysis*. Here we combine points with the cluster whose centroid or weighted average point is closest to the point to be joined. The centroid is recalculated at each stage. Centroid clustering has some disadvantages as well. When clusters are of unequal sizes, if one of the clusters is large, the new point to be fused is likely to be closer to the larger cluster than to the smaller ones. We might miss the characteristics of the smaller cluster.

Ward's method Ward (1963) proposed a solution to the problems associated with the techniques we have discussed so far. His solution is to combine new points to that cluster which will create the least information loss. 'Information loss' is defined as the error sum of squares. Ward's method does not compute distances between clusters. Rather it aims to reduce the total within-cluster sums of squares. If we were to assign the new point to any cluster, assignment to which cluster would result in the minimum increase to the total and within sum of squares? It is to this cluster that the new point should be assigned.

These are not the only hierarchical clustering procedures in use, although they are probably the most frequently used.

Non-hierarchical clustering (k-means clustering) In non-hierarchical clustering, each individual is assigned to one of k clusters. So (unlike in hierarchical clustering) the investigator must know the number of clusters in advance. All non-hierarchical methods use the following sequence: select k initial cluster centroids or seeds, assign each individual to the nearest centroid, and reassign each individual to k clusters using a predetermined stopping rule.

Seeds are selected using several procedures such as: choosing at random k observations with non-missing data; choosing the first k observations with non-missing data; and choosing the first observation with non-missing data as the first seed and then creating other seeds such that each is at least a predetermined distance from those previously selected.

Assignment of individuals to groups may be refined through reassignment. This can be accomplished using statistical or heuristic rules. The purpose of these rules is to ensure that the groups are as far from each other as possible. Some such rules are:

1. Reassign the individual such that it satisfies some relevant statistical criterion. Hill-climbing procedures in which some statistical criterion is minimized (such as the trace or the determinant of the within-group sum of squares and cross products matrix $\mathbf{W}$, or the largest eigenvalue or the trace of $\mathbf{W}^{-1}\mathbf{B}$, where $\mathbf{B}$ is the between-groups sum of squares and cross products matrix) fall under this group of techniques.
2. Calculate the centroid of each cluster and reassign individuals to cluster with the nearest centroid. For each assignment, recalculate the centroid of the group to which the individual is assigned as well the centroid of the group from which the individual is moved. Continue this process until the change in the centroids is less than the selected convergence criterion.
3. Assign all individuals to the nearest centroids, without recalculating them with each assignment. If the final centroids have changed in excess of a selected convergence criterion, then take another pass at reassignment. Continue this process until the change between the initial and final centroids is within the selected convergence criterion.

Which clustering method to choose

Hierarchical methods are not automatically superior to the non-hierarchical method or vice versa. Both methods have strengths and weaknesses and the method is chosen depending on the context.

Hierarchical methods

The main advantage of hierarchical methods is that they do not require the analyst to specify beforehand the number of clusters. Hierarchical clusters are often easier to interpret since what precedes and what follows clusters at each level can be seen clearly. When there are only a few entities such as computer models, they provide an easy-to-understand visual representation.

However, once an individual is assigned to a cluster, that individual may not be reassigned to a different cluster. Also, hierarchical grouping is susceptible to the linking effect, with a single entity having the power to decide the group memberships of many individuals. This linking problem is more pronounced with the nearest-neighbour than with the farthest-neighbour method.

Hierarchical techniques can also become computation-intensive as the number of individuals to be grouped increases.

Non-hierarchical methods

Traditionally, non-hierarchical methods are preferred in market segmentation analysis (a common application of cluster analysis in market research). These techniques require the analyst to specify *prior to the analysis* the number of clusters. Since there is no realistic way of knowing how many clusters to retain (let alone how many there 'are'), the analyst usually derives a number of alternatives – for instance, solutions for 2–10 clusters. After looking through several solutions, the analyst or the marketer might conclude that the four-cluster solution makes the most sense and is actionable from a marketing perspective and use this solution for further analysis of the market. (Although there are some heuristic or statistical criteria that may be used to identify the number of clusters, typically these criteria have not been found very useful in practice and therefore are not generally used in marketing research.)

Simulation studies indicate that randomly choosing the seeds results in poor partitioning. However, when hierarchical clustering is used prior to doing non-hierarchical clustering and the seeds are selected on the basis of this prior clustering as input to non-hierarchical clustering, the *k*-means clustering method provides superior solutions.

Viewed from this perspective, the two methods complement rather than compete with each other. Where they do not complement each other, the choice would obviously depend on the suitability of the problem to hand. For instance, if the objective is to see how different brands and products are grouped at different levels of abstraction or generalization, hierarchical clustering might be appropriate. To identify patterns in large amounts of data, on the other hand, may require the application of non-hierarchical clustering.

4.3 Cluster analysis: computer output

Hierarchical clustering

In the following example, consumers rated 15 models of car on the importance of eight attributes on a 10-point scale (higher numbers indicate more of the attribute): fun and playful, young and youthful, cheap and easy to maintain, strong brand reputation, flawless bodywork and finish, luxury, safety, and being just functional. The purpose of the analysis is to identify how the cars were grouped by consumers in terms of these eight attributes. A hierarchical clustering procedure using the single-linkage (nearest neighbour) method was applied to the data. The annotated output is given in Exhibit 4.3.

EXHIBIT 4.3
_____

Hierachical clustering output, annotated

Original data describing the characteristics of 11 car models

	Importance: Fun and playful	Importance: Young, youthful	Importance: Cheap and easy to maintain	Importance: Strong brand reputation	Importance: Flawless bodywork and finish	Importance: Luxurious	Importance: Safe	I consider my vehicle as just basic transportation
Ford cars	4.7	4.4	7.6	8.4	7.6	5.0	8.5	5.1
Chevrolet cars	4.6	4.2	7.6	8.5	7.7	5.2	8.6	5.1
Cadillac	4.7	4.4	6.5	9.0	8.7	8.6	9.1	3.9
Dodge cars	4.5	3.9	7.8	8.2	7.3	4.9	8.4	5.2
Daewoo	4.7	3.9	7.9	6.7	7.1	4.7	8.1	5.5
Honda	4.4	4.1	7.8	9.0	7.7	4.9	8.6	5.1
Acura	5.4	4.8	7.2	9.0	8.6	7.1	8.5	3.9
Hyundai	4.6	4.1	8.0	7.6	7.1	4.3	8.2	5.2
KIA	4.3	3.6	8.3	7.3	6.6	3.7	8.1	5.3
Infiniti	5.1	5.2	7.0	8.6	8.3	7.6	8.5	4.4
Toyota cars	4.1	3.9	7.8	9.0	7.6	4.8	8.6	5.4
Lexus	5.0	4.9	6.5	9.1	8.9	8.3	9.0	3.9
Audi	5.9	5.4	6.2	8.6	8.4	7.0	8.4	3.4
BMW	6.3	6.0	6.1	9.0	8.7	7.6	8.7	3.8
Mercedes	5.1	5.1	6.3	9.3	8.9	7.9	9.1	4.0

Squared Euclidean distance is computed as a measure of dissimilarity from the above table. This is labelled as the proximity matrix below.

Proximity Matrix

Case	1	2	3	4	5	6	7	8	9	10	11	12	13	14	15
1: Ford cars	0	0	18	0	4	1	8	2	5	9	1	16	12	17	15
2: Chevrolet cars	0	0	16	1	4	0	7	2	6	8	1	15	12	17	14
3: Cadillac	18	16	0	20	29	18	4	28	38	3	20	0	6	6	1
4: Dodge cars	0	1	20	0	3	1	11	1	3	12	1	20	16	22	19
5: Daewoo	4	4	29	3	0	6	18	1	2	17	6	28	22	30	28
6: Honda	1	0	18	1	6	0	9	3	6	11	0	17	14	20	16
7: Acura	8	7	4	11	18	9	0	16	24	1	11	2	2	4	2
8: Hyundai	2	2	28	1	1	3	16	0	1	17	3	27	20	28	25
9: KIA	5	6	38	3	2	6	24	1	0	26	6	38	30	40	37
10: Infiniti	9	8	3	12	17	11	1	17	26	0	13	2	3	4	2
11: Toyota cars	1	1	20	1	6	0	11	3	6	13	0	20	18	24	18
12: Lexus	16	15	0	20	28	17	2	27	38	2	20	0	4	4	0
13: Audi	12	12	6	16	22	14	2	20	30	3	18	4	0	1	3
14: BMW	17	17	6	22	30	20	4	28	40	4	24	4	1	0	3
15: Mercedes	15	14	1	19	28	16	2	25	37	2	18	0	3	3	0

This is a dissimilarity matrix

Based on the above dissimilarity matrix, and using the average linkage procedure, the following dendrogram was created.

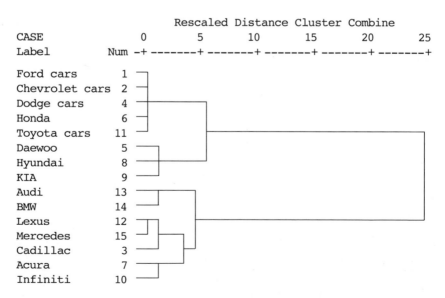

```
Dendrogram using Average Linkinage (Between Groups)

                        Rescaled Distance Cluster Combine
     CASE           0        5        10       15       20       25
     Label      Num -+ -------+ -------+ -------+ -------+ -------+

   Ford cars      1 ─┐
   Chevrolet cars 2 ─┤
   Dodge cars     4 ─┤
   Honda          6 ─┤
   Toyota cars   11 ─┘
   Daewoo         5 ─┐
   Hyundai        8 ─┤
   KIA            9 ─┘
   Audi          13 ─┐
   BMW           14 ─┤
   Lexus         12 ─┐
   Mercedes      15 ─┤
   Cadillac       3 ─┘
   Acura          7 ─┐
   Infiniti      10 ─┘
```

The results are fairly self-explanatory: There are two distinct groups of cars: *popular* (Ford, Chevrolet, Dodge, Honda, Toyota, Daewoo, Hyundai, and KIA) and *luxury* (Audi, BMW, Lexus, Mercedes, Cadillac. Acura, and Infiniti). Popular cars then split into two groups: the *Big 5* (Ford, Chevrolet, Dodge, Honda, Toyota) and *smaller brands* (Daewoo, Hyundai, and KIA). Premium brands in turn can be split into three groups: sports luxury (Audi, BMW), large luxury (Lexus, Mercedes, Cadillac) and entry-level luxury (Acura, Infiniti).

k-means clustering

In the study that follows, 972 salespeople were interviewed. They were asked to rate the importance of 25 different attributes on a 10-point numeric scale, where 10 meant 'extremely important' and 1 meant 'not at all important'. These ratings were submitted to a clustering program that uses the Howard–Harris algorithm (Neal, 1989). The purpose of the analysis is to identify salespeople segments, so they can be provided with the tools they need to sell. As is usually done in marketing research analysis, a number of alternative solutions (in this case with 2–10 clusters) were derived. On the basis of comparing different solutions from a marketing perspective, the analyst decided to use a five-cluster solution. The output is shown in Exhibit 4.4.

EXHIBIT 4.4

k-means clustering output, annotated

TABLE OF MEANS RANKED BY *F*-STATISTICS FOR FIVE-SEGMENT SOLUTION

The header below shows how many people fell into different segments. Some segments were small (segments 1, 111 and V) and others were larger (11 and 1V).

Subsequent lines show, for each variable, the overall mean and segment means. The last column shows the F ratio. Since the F ratio is an indication of a variable's ability to distinguish among segments, the variables are ranked according to the F ratio.

To understand what these clusters or segments are, we first identify the high and low values for each variable.

		SEGMENTS					
	TOTAL	I	II	III	IV	V	
0 SAMPLE SIZE	972	70	308	152	318	124	
							F-VALUE
3. Strong presence in loc mkt.	6.84	6.93	**8.38**	4.08	7.55	4.56	227.78
21. Cross-sell tools	7.42	3.20	**8.96**	6.71	**7.75**	6.02	191.65
8. Innovator in prdts/services	8.08	7.23	**9.27**	8.40	7.68	6.22	161.12
13. Training of complex prdts.	8.03	5.17	**9.36**	8.33	7.95	6.23	152.12
16. Has specialists available	8.38	7.26	**9.46**	8.70	8.18	6.47	128.53
7. Offers customized solutions	8.04	7.50	**9.17**	8.43	7.87	5.55	125.71
5. Innovator in technology	7.84	7.61	**9.04**	7.92	7.52	5.70	117.59
17. Full ste of electronic comm.	7.84	7.51	**9.01**	7.91	7.66	5.47	116.28

6. Brand name you trust	8.27	7.46	**9.39**	7.43	8.39	6.67	108.45
29. Brand name of the insurer	7.50	5.91	8.78	6.15	7.78	6.19	103.36
26. New product innovation	7.95	6.77	9.06	7.91	7.64	6.71	101.61
2. Leader in fin. serv.	7.93	7.31	9.07	6.95	8.00	6.45	98.86
4. Commitment to product excel.	8.76	8.53	9.57	8.83	8.48	7.51	85.39
28. Compensation to you	8.28	5.79	9.28	8.39	8.04	7.64	82.60
24. Quality of company prdts	9.11	8.56	9.73	9.40	8.75	8.45	81.73
14. Demonstrates flexibility	8.71	7.85	9.53	9.02	8.41	7.55	77.13
15. Realiable service provider	9.25	8.94	9.79	9.58	8.96	8.39	73.94
10. Rep you can contact direct	9.18	8.84	9.79	9.36	9.12	7.80	71.40
23. Relationship with company	8.60	7.98	9.50	8.61	8.43	7.13	66.89
25. Performance of inv. prdts	8.54	6.72	9.33	8.36	8.39	8.22	54.04
12. Get accurate responses	9.42	9.34	9.85	9.66	9.23	8.62	51.36
27. Competitive pricing	8.70	7.83	9.40	8.74	8.40	8.14	49.85
11. Responds in timely manner	9.32	8.91	9.80	9.58	9.14	8.53	43.42
1. Fin. strength & stability	8.83	8.36	9.54	8.56	8.71	8.01	42.59
25. Performance of inv. prdts	8.54	7.62	9.33	8.37	8.34	8.22	54.80

SUMMARY STATISTICS FOR 5 SEGMENT SOLUTION

The following ANOVA tables provide the mean, the standard deviation and the standard error for each variable. Standard error is particularly helpful in deciding whether the differences in means are significantly different from each other.

					SEGMENTS				
				TOTAL	I	II	III	IV	V
VAR	ANOVA	SUMMARY	SEGMENT STATISTIC	N=972	70	308	152	318	124
1	F-VAL	42.59							
	SSQ-B	272.86							
	SSQ-W	1548.72	MEAN	8.83	8.36	9.54	8.56	8.71	8.01
	MSQ-B	68.22	STD. DEV.	1.37	1.27	.91	1.67	1.18	1.62
	MSQ-W	1.60	STD. ERROR	.04	.15	.05	.14	.07	.15
2	F-VAL	98.86							
	SSQ-B	849.86							
	SSQ-W	2078.20	MEAN	7.93	7.31	9.07	6.95	8.00	6.45
	MSQ-B	212.46	STD. DEV.	1.74	1.65	1.13	1.97	1.21	1.91
	MSQ-W	2.15	STD. ERROR	.06	.20	.06	.16	.07	.17
3	F-VAL	227.78							
	SSQ-B	2692.76							
	SSQ-W	2857.93	MEAN	6.84	6.93	8.38	4.08	7.55	4.56
	MSQ-B	673.19	STD. DEV.	2.39	1.95	1.70	1.94	1.30	2.20
	MSQ-W	2.96	STD. ERROR	.08	.23	.10	.16	.07	.20

```
 4  F-VAL      85.39
    SSQ-B     424.05
    SSQ-W    1200.52    MEAN          8.76  8.53  9.57  8.83  8.48  7.51
    MSQ-B     106.01    STD. DEV.     1.29  1.09   .79  1.21  1.08  1.67
    MSQ-W       1.24    STD.ERROR      .04   .13   .04   .10   .06   .15

 5  F-VAL     117.59
    SSQ-B    1047.35
    SSQ-W    2153.14    MEAN          7.84  7.61  9.04  7.92  7.52  5.70
    MSQ-B     261.84    STD. DEV.     1.82  1.47  1.13  1.72  1.40  2.10
    MSQ-W       2.23    STD. ERROR     .06   .18   .06   .14   .08   .19

 6  F-VAL     108.45
    SSQ-B     860.54
    SSQ-W    1918.26    MEAN          8.27  7.46  9.39  7.43  8.39  6.67
    MSQ-B     215.13    STD. DEV.     1.69  1.34   .94  1.88  1.14  2.17
    MSQ-W       1.98    STD. ERROR     .05   .16   .05   .15   .06   .20

 7  F-VAL     125.71
    SSQ-B    1213.17
    SSQ-W    2332.99    MEAN          8.04  7.50  9.17  8.43  7.87  5.55
    MSQ-B     303.29    STD. DEV.     1.91  1.99  1.19  1.51  1.47  2.21
    MSQ-W       2.41    STD. ERROR     .06   .24   .07   .12   .08   .20

 8  F-VAL     161.12
    SSQ-B     982.00
    SSQ-W    1473.45    MEAN          8.08  7.23  9.27  8.40  7.68  6.22
    MSQ-B     245.50    STD. DEV.     1.59  1.56   .93  1.26  1.14  1.78
    MSQ-W       1.52    STD. ERROR     .05   .19   .05   .10   .06   .16

 9  F-VAL      71.40
    SSQ-B     365.98
    SSQ-W    1239.25    MEAN          9.18  8.84  9.79  9.36  9.12  7.80
    MSQ-B      91.50    STD. DEV      1.29  1.58   .55  1.21  1.00  1.88
    MSQ-W       1.28    STD. ERROR     .04   .19   .03   .10   .06   .17

10  F-VAL      43.42
    SSQ-B     179.13
    SSQ-W     997.33    MEAN          9.32  8.91  9.80  9.58  9.14  8.53
    MSQ-B      44.78    STD. DEV.     1.10  1.65   .60   .97  1.00  1.40
    MSQ-W       1.03    STD. ERROR     .04   .20   .03   .08   .06   .13

11  F-VAL      51.36
    SSQ-B     157.47
    SSQ-W     741.23    MEAN          9.42  9.34  9.85  9.66  9.23  8.62
    MSQ-B      39.37    STD. DEV.      .96   .93   .47   .69    98  1.39
    MSQ-W        .77    STD. ERROR     .03   .11   .03   .06   .05   .13

12  F-VAL     152.12
    SSQ-B    1533.98
    SSQ-W    2437.88    MEAN          8.03  5.17  9.36  8.33  7.95  6.23
    MSQ-B     383.49    STD. DEV      2.02  2.59  1.17  1.51  1.39  2.22
    MSQ-W       2.52    STD. ERROR     .06   .31   .07   .12   .08   .20
```

13	F-VAL	77.13								
	SSQ-B	470.62								
	SSQ-W	1475.09	MEAN	8.71	7.85	9.53	9.02	8.41	7.55	
	MSQ-B	117.66	STD. DEV.	1.42	2.01	.96	1.14	1.16	1.55	
	MSQ-W	1.53	STD. ERROR	.05	.24	.05	.09	.06	.14	
14	F-VAL	73.94								
	SSQ-B	232.05								
	SSQ-W	758.74	MEAN	9.25	8.94	9.79	9.58	8.96	8.39	
	MSQ-B	58.01	STD. DEV.	1.01	.98	.52	.78	1.01	1.25	
	MSQ-W	.78	STD. ERROR	.03	.12	.03	.06	.06	.11	
15	F-VAL	28.53								
	SSQ-B	929.02								
	SSQ-W	1747.35	MEAN	8.38	7.26	9.46	8.70	8.18	6.47	
	MSQ-B	232.26	STD. DEV.	1.66	1.74	.95	1.43	1.22	1.97	
	MSQ-W	1.81	STD. ERROR	.05	.21	.05	.12	.07	.18	
16	F-VAL	116.28								
	SSQ-B	1132.04								
	SSQ-W	2353.64	MEAN	7.84	7.51	9.01	7.91	7.66	5.47	
	MSQ-B	283.01	STD. DEV.	1.89	1.59	1.33	1.61	1.40	2.26	
	MSQ-W	2.43	STD. ERROR	.06	.19	.08	.13	.08	.20	
17	F-VAL	191.65								
	SSQ-B	2331.37								
	SSQ-W	2940.76	MEAN	7.42	3.20	8.96	6.71	7.75	6.02	
	MSQ-B	582.84	STD. DEV.	2.33	1.89	1.44	2.35	1.40	2.22	
	MSQ-W	3.04	STD. ERROR	.07	.23	.08	.19	.08	.20	
18	F-VAL	66.89								
	SSQ-B	550.90								
	SSQ-W	1990.90	MEAN	8.60	7.98	9.50	8.61	8.43	7.13	
	MSQ-B	137.72	STD. DEV.	1.62	1.41	1.13	1.64	1.27	2.11	
	MSQ-W	2.06	STD. ERROR	.05	.17	.06	.13	.07	.19	
19	F-VAL	81.73								
	SSQ-B	247.89								
	SSQ-W	733.27	MEAN	9.11	8.56	9.73	9.40	8.75	8.45	
	MSQ-B	61.97	STD. DEV.	1.01	.85	.58	.82	.95	1.25	
	MSQ-W	.76	STD. ERROR	.03	.10	.03	.07	.05	.11	
20	F-VAL	54.04								
	SSQ-B	450.73								
	SSQ-W	2016.30	MEAN	8.54	6.72	9.33	8.36	8.39	8.22	
	MSQ-B	112.68	STD. DEV.	1.59	2.55	1.04	1.83	1.18	1.51	
	MSQ-W	2.09	STD. ERROR	.05	.31	.06	.15	.07	.14	
21	F-VAL	101.61								
	SSQ-B	698.92								
	SSQ-W	1662.88	MEAN	7.95	6.77	9.06	7.91	7.64	6.71	
	MSQ-B	174.73	STD. DEV.	1.56	1.38	1.18	1.46	1.15	1.73	
	MSQ-W	1.72	STD. ERROR	.05	.17	.07	.12	.06	.16	

22	F-VAL	49.85							
	SSQ–B	270.54							
	SSQ–W	1311.89	MEAN	8.70	7.83	9.40	8.74	8.40	8.14
	MSQ–B	67.64	STD. DEV.	1.28	1.46	1.01	1.34	1.12	1.21
	MSQ–W	1.36	STD. ERROR	.04	.18	.06	.11	.06	.11
23	F-VAL	82.60							
	SSQ–B	813.84							
	SSQ–W	2381.78	MEAN	8.28	5.79	9.28	8.39	8.04	7.64
	MSQ–B	203.46	STD. DEV.	1.81	2.49	1.13	1.78	1.51	1.70
	MSQ–W	2.46	STD. ERROR	.06	.30	.06	.14	.08	.15
24	F-VAL	103.36							
	SSQ–B	1197.33							
	SSQ–W	2800.40	MEAN	7.50	5.91	8.78	6.15	7.78	6.19
	MSQ–B	299.33	STD. DEV.	2.03	2.01	1.45	2.17	1.34	2.21
	MSQ–W	2.90	STD. ERROR	.07	.24	.08	.18	.07	.20

DEGREES OF FREEDOM BETWEEN SEGMENTS = 4
DEGREES OF FREEDOM WITHIN SEGMENTS = 967

S U M M A R Y   T A B L E

The following table can be used to assess the reduction in variance as we increase the number of clusters. In marketing, no 'objective criterion' such as the reduction in variance is routinely used in identifying the number of clusters. Rather, the clusters that are accepted for the final analysis are based on their perceived relevance to marketing objectives.

0	NUMBER OF SEGMENTS	TOTAL WITHIN GROUP SUM OF SQUARES	AVERAGE VARIABLE VARIANCE PER GROUP	LAMBDA
0	1	63056.29	.000	1.00000
0	2	50235.53	2.238	.79668
0	3	46579.02	2.261	.73869
0	4	44608.23	2.172	.70744
0	5	43151.89	2.270	.68434
0	6	41929.94	2.196	.66496
0	7	40963.36	2.256	.64963
0	8	40341.11	2.134	.63976
0	9	39609.01	2.112	.62815
0	10	39041.46	2.080	.61915

Sometimes it is useful to track how a new cluster is formed. For instance, when we go from two-cluster to three-cluster solution, how was the third cluster formed? The following output (shown only for the first few respondents) identifies how each individual moved/stayed as different clusters were formed. This output can be summarized into 'migration tables' to understand the newly derived clusters in terms of earlier clusters.

CASE SPLIT	1	2	3	4	5	6	7	8	9
===============									
	2	3	3	3	3	3	3	3	3
	2	2	4	4	6	6	6	8	8
	2	3	3	3	3	3	3	3	3
	2	3	4	4	6	6	6	6	6
	2	2	2	2	6	6	2	6	6
	1	3	3	3	3	1	1	1	1
	1	3	3	3	3	1	1	1	1
	1	3	3	3	3	1	1	1	1
	2	3	4	4	6	3	6	6	6
	2	2	4	4	6	6	8	8	8
	2	3	4	4	6	6	6	6	8
	2	2	2	2	2	2	2	2	2
	1	3	3	3	3	7	7	7	7
	2	3	4	4	6	6	6	6	6
	2	3	4	4	6	6	6	6	6
	2	2	2	2	2	2	2	2	2
	1	1	1	5	5	7	7	7	7
	1	1	1	5	5	5	5	5	5
	1	3	4	4	4	4	4	4	4
	1	3	3	3	4	4	4	9	4
	1	1	1	5	5	5	5	5	5
	2	3	4	4	6	6	6	6	6
	1	3	4	4	4	4	4	9	10
	2	2	2	2	2	2	2	2	2
	1	3	4	4	4	4	4	9	10
	1	1	1	1	1	4	4	4	4

4.4 Cluster analysis: marketing applications

In this section, I present three applications of cluster analysis to marketing problems. They illustrate the variety of seemingly unrelated problems to which factor analysis may be applied.

How to design appealing travel packages

Marketing problem

The socio-economic and demographic characteristics of consumers do not always predict buyer behaviour effectively, especially in industries such as tourism. People with diverse demographic traits might show a similar tourism pattern; conversely, those with similar demographic traits might show a diverse tourism behaviour. A marketer who is interested in predicting consumer behaviour should obtain detailed knowledge of consumers by understanding consumer lifestyles. With this in mind, Ana González and Laurentino Bello (2002) segmented the Spanish tourism market based on consumers' lifestyles. The investigators' aim was to reveal the latent structure of the market in relation to individuals' lifestyles and thus to discover the specific features of each group and the relationship between lifestyle and tourist behaviour.

Application of cluster analysis

Four hundred people aged over 15 years who are resident in provincial capitals with a population greater than 100 000 in one autonomous community in Spain participated in the study. The respondents were selected using a multi-stage stratified random procedure. The content of the questionnaire is shown in Exhibit 4.5.

EXHIBIT 4.5

Questionnaire content

1. TOURISM (multiple-answer questions/nominal variables)
Short and long term: variables including destination, type of accommodation, means of transport, company on trip and reason for the choice of destination

2. LIFESTYLE (Likert scale of five points/ordinal variables)
Interests and opinions: questions that relate to society, politics, job and home milieu, personal success factors, company, environment, religion, future, family, friendship, responsibility, aspirations, attitude to personal problems, saving, innovation, and fashion

3. LEISURE ACTIVITIES (multiple-answer questions/nominal variables)
Various leisure activities to include DIY, sport, cinema, cultural activities, visits to beautiful places, joining social and religious associations, board games, night life, shopping, reading, music, TV programmes and radio programmes

Demographic questions and leisure activities (such as destination, watch TV, go to cinema, etc.) were measured on a nominal scale. Lifestyle attributes (such as 'I like saving regularly', 'Working alone, at home or elsewhere, is best', and 'I like trying new, different things') were measured on a five-point Likert scale (1, strongly agree; 2, agree; 3, neither agree not disagree; 4, disagree; and 5, strongly disagree).

González and Bello applied a variety of multivariate techniques to the data. Using principal components analysis, they reduced 43 lifestyle items to 23 factors and 110 activities items to 39 factors. Statistical analyses showed that these factors were correlated with travel behaviour. To further understand consumer tourism behaviour, the investigators applied k-means clustering to all the lifestyle factor scores derived as explained above. After examining a number of alternative solutions, González and Bello decided on a five-cluster solution that tested for the lowest association among variables across different clusters and highest association among variables within each cluster. This classification was later validated by discriminant analysis with a hit rate of 94% (see Chapter 6).

Results of cluster analysis

By examining how consumers in each segment responded to various items on the questionnaire, González and Bello were able to identify the nature of each of the five segments:

Home-loving lifestyle (**6% of the population**). This segment is made up of individuals fundamentally focused on family life. They attempt to enjoy a quiet and happy private life, and place emphasis on having children and the responsibilities that go with it. They are conservative and inflexible. They are cautious towards the future and take steps to protect themselves. For them, quality is more important than price and they are more demanding than those in other segments. Home-loving people enjoy cultural activities, such as visits to exhibitions, monuments or places of natural beauty. They enjoy a wide variety of reading material which may range from fashion to gossip. When they watch TV, they prefer news, documentaries, current affairs, debates and travel programmes. When they listen to the radio they also prefer news programmes. However, they have a strong aversion to sport in any mass media.

They make the fewest short trips and the most long trips. They are creatures of habit and go to the same destination year after year.

Idealistic lifestyle (**13%**). Consumers in this segment are interested in a better world and will fight against injustice. They have a desire to be in a job about which they can be enthusiastic. So they tend to collaborate in the enterprises employing them, and yet they are not interested in taking on managerial responsibilities. The idealists are flexible, responsible and tolerant in all matters. As consumers, besides giving quality priority over price, they are the most innovative group. They enjoy sport and music, especially classical. They go to concerts, theatre and dances. They are readers of magazines on politics as well as local and national papers. They dislike TV gossip programmes but like sports broadcasts. On radio, they prefer music channels and avoid game shows and comedies.

For short trips, this group chooses mainly inland destinations, making the second greatest number of visits to rural zones. This segment is unwilling to spend much money on weekend trips, and so stays with relatives or friends, or in guest houses. For long trips, this group travels mostly within the country and is fond of country villages, particularly those near their place of residence. The typical journey is with family and lasts for a week.

Autonomous lifestyle (**12%**). For this segment, success is linked fundamentally with individual freedom and independence. Members of this group place great emphasis on enjoying life. They work just to earn a living and they aspire to upward social mobility. Their political, religious and social views are liberal. They accept current social reality, and expect that the future will bring stability or improvement. They like the cinema, nightclubs, music (pop, rock, disco and ballads) and cultural activities such as exhibitions or touring monuments. They read sports and car magazines but not newspapers. On television they mostly watch films and on the radio they listen to sports programmes and current affairs.

This segment accounts for the largest number of weekend and long weekend or public holiday trips, mostly to city destinations, whether in Spain or outside. They stay in hotels and travel with friends. For longer trips, they go to coastal areas, especially in small villages. They go with friends for a one-week stay, using low-priced accommodation, while avoiding the homes of family or friends.

Hedonistic lifestyle (**36%**). This segment values human relationships and work as symbols of success. Yet the hedonists of this segment are not attracted to managerial positions. They are tolerant and take life as it comes and enjoy it. They try new products and have ecological leanings. Hedonists listen to music, read national and local newspapers, professional and business magazines, but not those related to fashion and the home. They are interested in news and films but are not enthusiastic about other media.

For short journeys, this group – more than any other – selects large cities, whether inland or on the coast. Members of this group travel in the company of friends. They use chiefly high-class hotels. For long journeys, this group selects destinations outside Spain, principally in Europe. When traveling within Spain, this segment prefers major cities, provincial capitals and coastal towns. The length of their holiday is usually two to three weeks, with accommodation in medium- to high-category hotels, apartments or serviced flats.

Conservative lifestyle (**33%**). This home-loving segment is focused on the well-being of the family, even when it is not fulfilling its expectations. When members of this segment have a problem, they turn to family or friends. Their jobs are not stimulating, and they work because they have to. However this does not bother them since they care more for working efficiently. While members of this highly materialistic segment aim to achieve managerial positions and upward social mobility, they seldom achieve it. They are religious but not tolerant, and are strict about law and order. They enjoy visits to areas of outstanding beauty, but dislike nightlife, modern music and the cinema. They are keen on both television and radio. On TV, they prefer to watch local news, reality shows, game shows and gossip programmes. As for radio, they listen to news, sports and humour.

This segment undertakes few weekend trips. Its members choose destinations that are principally within their own region, especially rural areas. They travel with their family and friends. For longer journeys, they prefer traditional domestic seaside destinations, especially country villages on the coast. They usually stay with family or with friends or in a house they own themselves or rent. The most frequent length of their holidays is from one to two weeks, spent together with their immediate family.

How cluster analysis addressed the problem

Cluster analysis identified the type of holidays that would appeal to different segments and how to appeal to each segment. For instance, the home-loving segment is interested in enjoying holiday arrangements corresponding to the needs of the family as a whole. Packages proposed by travel firms for this group should involve a longer stay (two weeks) in the summer to enable all family members to travel together. Destinations should be mostly by the sea, near the normal place of residence, with a mild and pleasant climate, and at a reasonable price. Tourist apartments might be the first choice, as these would allow the family unit to remain together for fairly long periods in conditions quite like those at home. Organized trips will not appeal to this group as they are likely to organize their own activities. Thus, travel operators could use direct marketing campaigns, through mailings, backed up in other mass media, principally radio and television advertising around the times of news programmes, to draw this group's attention to the existence of other alternatives. With regard to print media, regional newspapers and home and fashion magazines would seem the most appropriate. As the message would be connected to comfort for the family, enjoyed in a tranquil spot, it could be adapted to youngsters as well as to adults.

By contrast, tour operators targeting the idealistic segment would have the greatest likelihood of success by offering a wide range of countryside holidays. The package could include a combination of destinations, including interesting villages where consumers could stay with local residents and in towns where there are cultural activities. These holidays should include options allowing a chance to take an active part in rural activities, on the one hand, and, on the other, giving access to various sports activities. The packages could be aimed at small groups, made up of a family and a few friends, and with a duration of eight to ten days. Any media could be used, but radio and television advertisements during sports and music programmes would be best, while provincial and national newspapers might be the most effective channel.

The use of cluster analysis allowed González and Bello to draw similar conclusions (in considerably greater detail than indicated here) for each of the five segments.

How to deliver advertising messages to seniors

Marketing problem

Seniors in the United States outnumber the combined populations of Canada, Israel, Switzerland, and Ireland. This market is growing in size and is financially potent, controlling a disproportionate per-capita share of discretionary income in the United States. Most marketers recognize the heterogeneity of this group and try to deal with its diversity. While demographic segmentation approaches are useful for many marketing decision areas, they often fall far short of the ideal for the purposes of planning advertising and provide limited insight into creative approaches and communication planning. An important issue for advertisers targeting the aged consumer is the way in which the elderly utilize and evaluate information from advertising in making purchase decisions. Although elderly consumers have been segmented in this regard, Davis and French (1989) sought to address some issues specifically: to identify potential audience segments; to develop psychographic profiles for each of these potential segments; and to examine media consumption among these segments.

Application of cluster analysis

Davis and French used the data generated by the annual lifestyle surveys sponsored by DDB/Needham Advertising. The sample, drawn from a consumer panel, consisted of 217 married females, not employed outside the home, and at least 60 years of age. Because the respondents possessed somewhat higher education, income, and activity levels than those found in the general elderly population, the findings may not be generalized to the elderly population as a whole.

A number of psychographic statements in the survey dealt with attitudes, interests and opinions. Respondents were asked to rate their degree of agreement with a number of statements dealing with psychographics on a six-point scale. Respondents were also asked to rate their use of various mass media. The study was replicated by collecting identical information from a separate sample of 232 females aged 60 and over who were part of an earlier lifestyle study.

To segment the market according to information usage and beliefs about advertising, four statements were used:

- I often seek out the advice of friends regarding brands and products.
- Information from advertising helps me make better buying decisions.
- I don't believe a company's ad when it claims test results show its product to be better than competitive products.
- Advertising insults my intelligence.

Ratings on these four statements were used as input to cluster analysis using the Ward method. (The Ward clustering method employed to analyse data from the replication sample produced roughly similar results.) Based on the dendrogram generated by cluster analysis, Davis and French identified three natural clusters.

Results of cluster analysis

By reviewing the means of the four statements for each cluster (Exhibit 4.6) the investigators named the segments as 'engaged', 'autonomous' and 'receptive'.

EXHIBIT 4.6

Key variables by segment

| | Segment | | |
Variable	Engaged	Autonomous	Receptive
Advertising insults my intelligence	5.24	4.86	2.20*
Information from advertising helps me make better buying decisions	4.69	3.65	4.78
I often seek out the advice of friends regarding brands/products	4.55	2.16*	2.99*
I don't believe a company's ad when it claims test results show its product to be better than competitive products	4.78	4.85	4.12

* Disagree with the statement

- *Engaged* (25% of the market). The engaged segment is highly involved in social activities. Members of this segment generally agree that advertising insults their intelligence and do not believe an advertiser's test results showing its product to be better than competing products. The engaged segment is sceptical of advertising and seems to rely extensively on external information sources.
- *Autonomous* (41% of the market). The autonomous segment does not seek out the advice of friends regarding brands and products and is neutral about the value of information from advertising in making purchase decisions. Members of this segment feel that advertising insults their intelligence and are highly suspicious of competitive advertisements. They do not seem to rely heavily on any external sources for information. Rather, they rely heavily on personal experience as an internal information source.
- *Receptive* (34% of the market). This segment, unlike the other two, does not believe that advertising insults its intelligence and is receptive to advertising communications. While members of this segment are suspicious of competitive advertising, they believe that advertising helps them make better buying decisions. They tend to utilize and rely on information from advertising, and tend not to consult with friends about brands and products. The attitudes these women have towards advertising would make this group a more receptive audience for advertisers, especially compared to the other two groups.

To further understand the segments, four factors (which can roughly be thought of as 'cosmopolitan', 'cooking and baking', 'innovative and concerned', and 'negative business outlook'[1]) were derived from a shortlist of 41 lifestyle/psychographic statements included in the questionnaire. Exhibit 4.7 shows the factor scores for each

---

[1]While the authors suggested what these factors might be, they did not name them, as is done here.

segment, along with the means of variables that loaded highly on these factors. This information was used to provide psychographic profiles for each of the three groups identified through cluster analysis of the study sample.

EXHIBIT 4.7

Market segments vs. lifestyle factors*

Lifestyle factors		Market segments		
Factor and Variables	Loading	Engaged	Autonomous	Receptive
Factor 1: Cosmopolitan				
I am interested in the cultures of other countries	0.59	4.4	3.9	3.9
I get personal satisfaction from using cosmetics	0.48	4.3	3.7	3.5
I enjoy looking through fashion magazines	0.42	4.9	4.3	4.6
Factor 2: Cooking and baking				
I like to bake	0.70	5.5	4.8	5.2
I like to cook	0.61	5.3	4.6	5.0
I always bake from scratch	0.54	3.8	3.2	3.6
Factor 3: Innovative and concerned				
I try to select foods fortified with vitamins/minerals	0.49	4.9	4.4	4.6
I try to buy a co.'s products that support educational TV	0.43	4.1	3.5	3.7
I am usually among the first to try new products	0.43	3.5	2.8	3.2
Factor 4: Negative business outlook				
Manufacturers' warranties are not worth much**	0.50	3.3	3.5	2.8
Most big companies are just out for themselves	0.48	4.3	4.5	3.9
TV advertising is condescending toward women	0.41	4.3	4.2	3.6

* Factor names are not given, but suggested by the authors. Some variable names are shortened
** Actual wording: Generally, manufacturers' warranties are not worth the paper they are printed on

   Members of the engaged segment scored high on the cosmopolitan factor. They are interested in the cultures of other countries and receive personal satisfaction from using cosmetics and looking through fashion magazines. Cooking and baking seem to be an important part of their lifestyle. They display personal and social concerns but are not innovative, although they are more inclined to try new products compared to others. They display negativism, believe that TV advertising is condescending towards women and that most big companies are just out for themselves, and tend to be suspicious about the value of manufacturers' warranties.

Those in the autonomous segment scored lower on the cosmopolitan dimension. They enjoy looking through fashion magazines, using cosmetics and cooking and baking, but less so than those in the engaged segment. They are less concerned about educational television and vitamin consumption and are the least innovative of the three groups. They display a moderate amount of negativism.

Receptive segment members score very low on the cosmopolitan factor. They are less interested in other cultures and in the use of personal cosmetics compared to those in other groups. Their interest in cooking and baking is moderate but they score high on the innovativeness-social issues factor. Educational television and nutrition interest them. They score low on the negativism factor.

These three segments also exhibited different media consumption habits. Exhibit 4.8 summarizes the differences.

EXHIBIT 4.8

Levels of media consumption by segments

Segment	Media	Consumption
Engaged	Levels of newspaper readership	High
	Viewing levels for television news programming	High
Autonomous	Levels of newspaper readership	Moderate
	Viewing of television media in general	Low
Receptive	Levels of newspaper readership	Moderate
	Viewing levels of television comedy programs	High

How cluster analysis addressed the problem

The results of cluster analysis provided a number of insights to advertisers. For example:

- The engaged segment can be of special interest to companies introducing new products for elderly consumers, because it is more innovative. We can reach people in this segment through mass media, especially news media. Because they are socially involved and often consult with friends regarding products, new-product knowledge could be propagated through word-of-mouth communication. The negativism of those in the engaged segment creates special challenges to advertisers trying to communicate with this group. These women have many negative attitudes towards business in general, particularly advertising. Special attention should be paid to designing a message that is credible, yet not condescending. Comedic appeals may not work very well with this segment. The heavy emphasis they place on news programming suggests that a factual and serious approach might be most appropriate. Appeals that are well documented and educational in nature could be successful.
- People of the autonomous segment would not be as attractive as targets for new

products since they are the least innovative of the three segments. We may need to find alternative ways of communicating with them since they seem to be socially isolated and do not use mass media as heavily as those in the engaged segment. Women in the autonomous segment appear to be quite similar to the women in the engaged group in their need to reinforce self-esteem. This segment also displays a high need for security. Appeals that work well with those in the engaged segment – such as well-documented appeals – would probably also work well with the autonomous segment.

• Members of the receptive segment are a particularly attractive target for businesses that rely on advertising to communicate with prospects. This segment has favourable attitudes toward advertising and business. Comedic appeals or advertising done in connection with television comedy programmes might be effective in reaching this group.

Because these three segments have different characteristics, we may have to communicate with them differently and perhaps in different media. Cluster analysis, by identifying the characteristics and media habits of different segments, provided some overall guidelines in this regard.

How to reach the affluent

The marketing problem
The number of US households with an annual income of $100 000 and over has been growing steadily for several years. The wealthiest 1% of American families account for approximately 11% of income. Since income is expected to relate to money spent on financial services (which might include investment advice, asset management, trusts, executorships, and a variety of loan products), affluent customers could be very profitable for financial institutions.

Stanley, Moschis and Danko (1987) hypothesized that affluent segments vary in their usage of financial services and the differences among segments can be explained by specific demographic and socio-economic variables. Such clustering can be carried out using many methods. In order to provide a knowledge base for the financial services industry and to researchers in this field, Stanley *et al.* carried out a cluster analysis of affluent customers.

Application of cluster analysis
After reviewing a number of alternatives, Stanley *et al.* concluded that attributes related to the usage of three broad categories of financial services are likely to yield the best results: (1) speculative (e.g., options, commodity contracts, and other types of high-risk securities) and highbrow investments (e.g., antiques, gems, and precious metals); (2) investment decision delegation services (e.g., hiring an investment manager to look after investments); and (3) credit and credit-related services (e.g., loans, credit cards and overdrafts).

Stanley *et al.* selected a random sample of census neighbourhoods or block groups from a listing of neighbourhoods with average incomes exceeding $50 000 in those states ranked in the top half by income according to 1980 census data. From this, they selected 9000 households at random for a mail survey. A completed questionnaire was returned by 42.3% of the valid addresses. Of these, 2914 qualified for segmentation analysis.

The survey contained questions about usage of 56 financial products and services. To derive segments from the usage data, Stanley *et al.* used hierarchical cluster analysis. The investigators used the hierarchical approach since 'it does not require any prior estimation of either the number of clusters expected or their probable characteristics'.

Results of cluster analysis

Stanley *et al.* derived seven segments on the basis of their cluster analysis. An examination of these clusters showed that each cluster has a unique pattern of service usage, as shown in Exhibit 4.9. The segments were named on the basis of the usage patterns shown in this exhibit.

EXHIBIT 4.9

Affluent market segment profiles

Segment size	Total	Speculator	Entrepreneur	Passive investor	Below normal	Retirement planner	Highbrow	Bridger
	100%	11%	17%	13%	17%	24%	15%	3%
Mean household income ($000s)	87	110	99	86	67	76	92	108
Mean net worth ($000s)	668	748	995	629	502	447	745	1072
% earning over 100K	25	42	34	24	12	19	23	35
Net worth >$1 000 000 (%)	13	18	24	15	7	5	14	24
Mean age of respondent	53	49	54	51	58	53	52	51
Age of respondent <45 yrs (%)	29	39	21	29	19	23	31	30
Business owners (%)	53	51	96	65	31	35	51	68
Investment as % income	16	17	17	16	15	14	17	15

The *speculator* segment accounts for 11%. Members of this segment are younger, have the highest average income and one in five in this segment is a millionaire. They score high along the speculative investments scale, have a fairly heavy credit orientation and use margin loans liberally. Forty per cent of this group are corporate executives, compared to 30% overall. This segment invests more (in terms of both dollars and proportion of income) than any other group. It also spends significantly more money on interest on loans than the average for the sample ($17 700 versus $13 000).

The *entrepreneur* group (17%) ranks first along the business dimension and last on speculative investments. Its use of services does not show any significant pattern. Entrepreneurs have a relatively high average income ($98 000) and net worth ($995 000). But, as a proportion of net worth, the entrepreneur has the lowest return (9.9%). A reason for this might be that closely held businesses have a great deal of discretion as to when to realize income, and they tend to realize lower rates of return on their wealth.

Entrepreneurs on average are heavy investors and credit users. They invest 17.3%

and spend 17.8% of their personal household income annually on servicing loans. Practically all (96%) in this segment are business owners. This finding would suggest that many affluent individuals who name their occupation as salaried employees are in fact entrepreneurs.

The *passive investor* segment (13%) is made up of investment delegators. They are less likely to use professional financial planning and hold fewer highbrow investments.

Passive investors are heavy users of prestige credit. They spend more of their income on interest on loans ($13 800 or 16.1%) than the typical respondent ($13 000 or 15%). This segment is quite typical of the entire sample with regard to the proportion of members who were in the various occupational categories.

Members of the *below normal* segment (17.3%) occupy a lower position than the typical affluent respondent in terms of the many usage areas: business/credit, traditional credit, prestige credit, financial manager, financial planning, and highbrow investments. They are somewhat average in regard to their usage of non-traditional credit and speculative investments. The below normal group is also below average with regard to all measures of affluence. It has the lowest average income ($67 000) and second lowest average net worth ($502 000). Only one in eight of its members is in the six-figure income category, and only one in 14 is a millionaire.

This segment is older (average age was 58 years) and contains a relatively small portion of younger and middle-aged respondents (19%). It contains the smallest proportion of entrepreneurs (31%) but a large proportion of inherited affluent (24%). The below average segment spends on average the fewest dollars on loan interest ($6800 or 10.2% of their personal household incomes), and invests fewer dollars ($10 200 on average). This is not surprising, considering that a large proportion of this segment consists of retired people.

The *retirement planner* segment is the largest (25%), and its members score average or below average on all but one scale, financial planning. They score the lowest on both the business credit and highbrow investment usage scales and the second lowest in the financial manager dimension. This economic underachiever group represents the lower economic end of the affluent continuum. It has the lowest average net worth ($447 000) and second lowest income ($76 000). Only one in 20 of its members is a millionaire. Consistent with their higher than average proportion of income to net worth (17%), only a third in this group own a business. They spend less on interest on loans compared to other segments. This segment demonstrates a considerable interest in retirement or defensive services, and is less likely to include offensive and aggressive wealth creators.

Members of the *highbrow* segment (15%) invest in artefacts that are often pretensions of wealth (e.g., antiques, precious gems, precious metals, and paintings). This segment resembles the total sample in terms of its position on all scales except for the business credit dimension, where it ranks lower than average. It contains the highest portion of inherited wealth respondents. Also, it is composed of a larger than expected proportion of middle-level managers, teacher/professors, and salespersons. They invest a significantly greater proportion of their income than the average for the sample and spend a small percentage on interest on loans. The highbrow group does not borrow heavily and is not a heavy user of investment management services.

The *bridgers* (3%) are something of an outlier group and rank high on the financial manager and financial planning dimensions. Members of this segment are relatively younger but have the highest average net worth (over $1 million) and are more often business owners. Nearly a quarter of the bridgers' realized annual household income, on average, is allocated to interest on loans. They invest a lower proportion of their income than the average respondent. The bridger is similar to the entrepreneur with regard to the proportion of income to net worth.

By definition, bridgers often are 'in transition'. They borrow considerable sums of money (short term) in anticipation of the sale of personal real estate, commercial real estate, and/or business assets. This segment contains a high portion of younger corporate executives. Nearly seven out of ten of the bridger segment are business owners.

How cluster analysis addressed the problem

The results of this study suggest that financial service usage is directly related to the level of affluence measured in terms of income and/or net worth. Consumers within affluent neighbourhoods may differ in terms of their sensitivity to promotional messages about financial services. Because seven affluent usage segments can be found within the same geographically defined affluent market, response to advertising is also likely to vary. Consequently, direct mail marketing methods may be more productive when a two-stage design is utilized. The first stage would be used to qualify by a mail promotional type of survey the affluent prospect in terms of level of wealth, interest in products, and promotional sensitivity parameters based on an estimate of their size and date of variable compensation. Based on this information, each prospect can be categorized in terms of estimated user type. In addition, during this second stage, the marketer can time his messages given the prospect's estimated financial situation. Also during this stage, any changes in the customer's situation can be noted.

Although some markets may have doubts about the affluent responding to the first stage of the method, Stanley *et al.* state that the affluent are very responsive to surveys that require more than an hour to complete.

In more specific terms, the results of cluster analysis can be used to solve many marketing problems, as illustrated in the following paragraphs.

The speculator segment is most often the primary target segment of security brokers. The results of the study, however, suggest that fewer than an eighth of affluent individuals are in this category. Also, most other affluent consumers are less likely to take financial risk and are more interested in long-term steady returns on their investments. Brokerage companies should pay more attention to the other segments as well – there are more affluent consumers interested in borrowing money than in investing money through a broker.

The bridger segment is the smallest of all segments. Yet its members spend more on interest per capita on loans than any other segment. Being a member of this cluster is often a temporary situation. Providing bridging loans can give the provider a base for the cross-selling of other services. Bridging loans typically require complete financial statements. These financial statements, where it is legal to use them, can provide a foundation for information-based cross-selling.

The passive wealth cluster contains a large proportion of affluent sales professionals. Sales professionals often prefer to delegate personal investment decisions. Although sales professionals do not make up the entire cluster, their presence in the affluent market deserves some attention. The presence of this market provides a significant and seemingly untapped opportunity. Affluent sales professionals are important candidates for various types of tax-advantaged planning programs as well as deferred compensation packages.

Analyses similar to the above can be used to understand and service affluent markets.

4.5   Caveats and concluding comments

Cluster analysis is not a strict mathematical technique. Rather it is a set of heuristic techniques that use mathematical formulae. Consequently, one gets different sets of clusters depending on the heuristics used. There will be alternative solutions to a problem that may be considered equally valid. Marketers tend to choose a solution that is likely to be useful from a marketing perspective or one that has worked in the past.

Different clustering techniques may give different solutions

As we noted, different techniques can and do give different solutions. No technique can automatically be considered superior to others. As is true of data analysis in general, stronger patterns are usually detected by almost any of the alternative techniques available, but techniques differ in their ability to detect weaker (or non-existent) patterns. Even when we use k-means clustering, the results of different procedures may give different results. If clusters are far apart and well differentiated, it would matter little whether we use the single-linkage, complete-linkage or Ward's method or some form of k-means clustering – all of the techniques are likely to identify well-differentiated clusters. However, as patterns get weaker, different techniques may give different solutions. While there are a number of 'validation procedures', there is no consensus as to which one is the best and, consequently, none of them is used widely in marketing research.

Clusters are 'imposed' rather than 'derived'

The objective of cluster analysis is to identify clusters, which implies 'identify clusters, if any in the data'. However, cluster analysis will always produce clusters whether there are any natural groupings in the data or not. Just because the technique produces clusters, we cannot assume that they are meaningful.

The number of clusters in the data is unknown

This is related to the previous problem. If the cluster solutions are 'imposed' by the analyst, it also follows that there is no clear answer to the question 'how many clusters?'

All the above problems are in some way a function of our inability to define what a cluster is. There is no precise mathematical definition of the term 'cluster', although there are several rules that will help us to derive this elusive identity.

This inevitably leads to the question whether it is worthwhile using this technique at all, given that we cannot define the term 'cluster' precisely, and that we do not know whether there are any clusters and, if there are, how many. Experience shows that, despite all its limitations, cluster analysis is useful in marketing analysis. Although the term 'cluster' itself is not precisely defined, there is wide agreement on the properties of clusters such as cluster variance. Eventually, for clusters to be useful in marketing, they must make intuitive sense. Consequently, solutions that are accepted in marketing are based on criteria such as:

1.   Do the clusters make logical sense?
2.   Do the clusters differentiate customers? (For instance, if our clusters are based on car buyers, do people in different segments prefer or use different types of cars?)
3.   Are the clusters stable? (Since market segments derived through cluster analysis

are used to plan marketing strategies, it is important that the clusters are stable and reproducible, even if imperfectly.)

Since market segmentation is one of the basic strategies used in marketing, cluster analysis which identifies such segments has found extensive applications in marketing. With all its limitations, cluster analysis is considered to be a useful technique, especially in identifying market segments.

Bibliography

Further reading

A lucid and comprehensive survey of cluster analytic techniques can be found in:

Everitt, Brian S., Landau, Sabine and Leese, Morven (2000) *Cluster Analysis* (4th edition). Arnold, London.

The following book provides an introduction to cluster analysis and introduces a variety of algorithms and computer programs:

Kaufman, Leonard and Rousseeuw, Peter J. (1990) *Finding Groups in Data*. Wiley Interscience, New York.

To understand how cluster analysis is used in marketing, you may want to refer to:

Wedel, Michel and Kamakura, Wagner (2000) *Market Segmentation: Conceptual and Methodological Foundations* (2nd edition). Kluwer Academic Publishers, Boston.

A non-technical introduction to market segmentation using cluster analysis and other techniques can be found in:

Myers, James H. (1996) *Segmentation and Positioning for Strategic Marketing Decisions*. American Marketing Association, Chicago.

References

Davis, Brian and French, Warren A. (1989) Exploring advertising usage segments among the aged. *Journal of Advertising Research*, **29**(1), 22–9.
González, Ana and Bello, Laurentino (2002). The construct 'lifestyle' in market segmentation: The behaviour of tourist consumers. *European Journal of Marketing*. **36**(1–2), 51–85.
Neal, William D. (1989) A comparison of 18 clustering algorithms generally available to the marketing research professional. In *Proceedings of the Sawtooth Conference*, Vol. 1. Sawtooth Software, Sequim, WA.
Stanley, Thomas J., Moschis, George P. and Danko, William D. (1987) Financial services segments: the seven faces of the affluent market. *Journal of Advertising Research*, **27**(4), 52–67.
Ward J. (1963) Hierarchical grouping to optimize an objective function. *Journal of American Statistical Association*, **58**, 236–44.

Part 3

Dependence Techniques

5

Multiple Regression Analysis

5.1   What is multiple regression analysis?

A marketer is often faced with the problem of identifying how a critical marketing variable is affected by a number of other variables. How do different attributes of an organization affect customer loyalty? How do variables such as inflation, discretionary income, and advertising expenditures affect sales? Which of a product's attributes influence the overall evaluation of that product? What is the contribution of different service attributes to customer satisfaction? What are the 'key drivers' of purchase behaviour? Answers to questions like these enable the marketer to deploy resources effectively and predict, however weakly, the course of future events and the potential effects of certain marketing actions.

Regression analysis, the most widely used of all statistical techniques, aims to identify how a set of variables (such as product quality, service quality and price) is related to another variable of interest (such as customer satisfaction). The variable of interest is commonly known as the *dependent* or the *criterion* variable, while the influencing variables are called the *independent* or the *predictor* variables. Regression analysis enables us to identify the influence of predictor variables both individually and collectively. Regression analysis answers the question 'what sets of weights attached to independent variables will predict the value of the dependent variable with maximum accuracy?' The criterion variable is seen as a linear (weighted) combination of predictor variables. Regression analysis is a metric technique. Since variables are not inherently dependent or independent, it is for the marketer to specify which variable is considered dependent and which are considered independent. (For instance, it may be considered equally valid to say that satisfaction with the price leads to satisfaction with the company, and that satisfaction with the company leads to satisfaction with the price.)

Here are some examples of marketing problems which attempt to relate a number of independent variables to a dependent variable, making them suitable for tackling by means of regression analysis:

1.   Customers rate a company on a 10-point scale on a number of characteristics such as customer service, price competitiveness, product quality, and responsive-

ness. The marketing manager would like to know whether these ratings will enable us to predict how a customer would rate the company overall. She would also like to know the exclusive effect of each one of these attributes on the dependent variable.

2. A luxury goods manufacturer would like to forecast their sales for the next 12 months. Past experience has shown that factors such as inflation rate, interest rate and discretionary income affect the purchase of luxury goods. The manufacturer has data for the past 20 years on these variables and would like to understand how they affect sales, so the relationship can be used to predict the sales for the next 12 months.

3. A marketer has a database with a wealth of information on customer demographics and a score for loyalty (self-rated intention to buy again). The marketer would like to know which of the demographics influence loyalty.

5.2   The regression analysis model

We start with a number of independent variables (such as customer ratings of different product attributes) and a single dependent variable (such as ratings on purchase intent). How do ratings on product attributes influence intent to purchase? We can use regression analysis to find weights for each of the independent variables that would best predict the purchase intent (rating) of the customer. Expressed formally, given the value of the dependent variable y and the values of independent variables x_1, x_2, x_3, ..., x_p, the task of regression analysis is to find values of coefficients b_1, b_2, b_3, b_4, ..., b_p that would best predict y. The linear regression equation is of the form

$$y' = b_0 + b_1 x_1 + b_2 x_2 + b_3 x_3 + b_4 x_4 + \dots + b_p x_p,$$

where x_1, ..., x_p are independent variables, b_1, ..., b_p are regression coefficients (weights attached to corresponding variables x_1, ..., x_p), y' is the predicted value of y, and b_0 is the value of y' when x_1, ..., $x_p = 0$.

The best-fitting regression coefficients are produced by the equation for which the errors of prediction are at a minimum. The 'errors of prediction' are defined as the sum of squared differences between the estimated values of the dependent variable (y') and the observed values (y). Because we minimize the squared errors of prediction $(y - y')^2$, this is known as the least-squares solution. If we let $e = y - y'$, we can describe the actual y values as

$$y_i = b_0 + b_1 x_1 + b_2 x_2 + b_3 x_3 + \dots + b_p x_p + e_i, \quad i = 1, \dots, n.$$

The deviations e_i are assumed to be independent of each other and normally distributed with 0 mean and constant variance.

The importance of attributes: beta coefficients

While the predictive equation

$$y' = b_0 + b_1 x_1 + b_2 x_2 + b_3 x_3 + b_4 x_4 + \dots + b_p x_p$$

enables us to predict the value of y, the size of the bs tells us nothing about the relative importance of different variables. The b coefficient of the variable x_1 can be twice as big as x_2, but it does not follow that x_1 is twice as important as x_2 since these variables could have been measured in different units. For instance, one variable could have been income and the other a rating on a 10-point scale. Even if two variables are measured on the same scale, the *b* coefficients may still be not comparable because one variable may have a higher variance than another. Consequently, to understand the relative importance of the variables, we need to convert the predictive equation to its standardized or *z*-score form (where each coefficient has a mean of 0 and standard deviation of 1),

$$z_{y'} = \tilde{\beta}_1 z_1 + \tilde{\beta}_2 z_2 + \tilde{\beta}_3 z_3 + \tilde{\beta}_4 z_4 + \dots + \tilde{\beta}_p z_p,$$

where $z_{y'}$ is the predicted standardized score on the dependent variable. The coefficients of the standardized predictor variables are called the *beta coefficients* or the *beta weights*.

Betas, also referred to as *partial correlation coefficients*, can be used to *order* the importance of independent variables in predicting the dependent variable. Because all betas are standardized on the same scale (mean of 0 and standard deviation of 1), it now is possible to compare the contribution of different variables since they are in the same units. While betas tell us about the relative importance of variables, they do not tell us anything about their absolute importance, since betas will change depending on other variables in the equation.

How well does the model fit?

An obvious way to assess a model is to assess how well a set of independent variables correlates with the dependent variable. Multiple correlation (R), which is the correlation between the dependent variable and the independent variables, is used in this context. The measure R^2, or the *coefficient of determination*, expresses the sum of squares due to regression as a proportion of the total sum of squares. Thus (multiplied by 100) it is the percentage of total variance explained by the model, and is an indicator of model fit:

$$R^2 = SS_{REG}/SS_T$$

The partial F statistic (which is the F ratio, net of the included variables) is given by

$$\text{Partial } F = (SS_{REG \text{ (variables in)}} - SS_{REG \text{ (variables out)}} / SS_{RES \text{ (variables in)}})$$

Selecting the best set of predictors

There are many methods of regression analysis that will enable us to identify the best set of independent variables. However, different methods may identify different sets of variables. Also, two different models may identify equal numbers of variables and may have the same R^2, and yet these variables may be different from model to model. A marketer who is interested in finding out the most critical service attributes that drive customer satisfaction can potentially get different answers, depending on the model.

Hierarchical methods

One method of identifying the best set of variables is known as *backward elimination*. We start with the full model of *p* variables, and eliminate the variable with the smallest partial F-value. In the resulting model we have $p-1$ variables, from which we eliminate the variable with the smallest partial F-value. We continue this process until some prespecified criterion (e.g., minimum F-value, level of significance) is met. Although this is a logical way of identifying the 'best predictors', there might be other combinations of variables that may provide as good a solution. Suppose we start with 10 independent variables and identify variables 3, 7, 8, and 10 to be the best predictors. It is quite conceivable that variables 3, 7, 4 and 9 would have done the job equally well.

In the *forward selection* procedure, the variable that has the highest F-value is chosen and a regression equation is created using just one independent variable. In the second step, the variable that, when added to the first, would maximize the overall F is included in the equation. This process is continued until a prespecified criterion is met.

Stepwise regression is similar to the forward selection process. However, in stepwise regression, whenever an additional variable is chosen, the partial F-values of all variables already in the equation are examined and any variable that falls below a prespecified criterion is eliminated and a new variable is included in its place.

Non-hierarchical methods

In *all-subsets* regression, all possible sets of predictive variables are entered. Although the availability of cheap computing power enables us to do this, it may not always be a good idea to substitute brute computing force for logical selection of variables, aided by computing to decide between well-chosen alternatives. All-subsets regression can also complicate the problem of modelling if several subsets fit the data equally well.

An alternative is simply to enter all independent variables together, especially if one has some past experience with the kind of data being analysed. Sometime such *simultaneous regression* is used to test theoretical models.

The 'best' method may depend on the context of analysis. As a general rule, it is preferable to start with mental models as well as hypotheses based on theory and past experience rather than solely depending on regression analysis to develop a model that would explain the dependent variable.

Using categorical independent variables

Although regression is a metric technique, it provides a means of including non-metric variables (such as gender and educational level) as predictors. The non-metric variables used in regression analysis are referred to as dummy variables. If there are n categories in a non-metric variable, $n - 1$ variables will be needed to specify them in the equation. For instance, gender has two categories so we need only one variable. If we let x stand for gender, we can give a value of 1 to male and 0 to female. Again, if income has three categories (low, medium and high), we need two variables x_1 and x_2 which can be coded as follows:

	x_1	x_2
Low income	1	0
Medium income	0	1
High income	0	0

Although this is the most common way of coding, other types of coding are also used in regression, depending on the theoretical model.

5.3   Regression analysis: computer output

In a survey, 1010 people who had no savings were asked how likely it was that they would start a savings programme in the next 12 months. They were also asked to rate how well they agreed with a number of statements:

- Concerned about rising cost of living
- Should save 10% of income
- Should establish retirement. Plan early
- Accept risk for better rate
- Cannot afford to save
- Believe in insurance
- Put money away regularly
- Save adequately for retirement
- Saving for retirement top priority.

What is a person's likelihood of saving, given the importance he/she attaches to these attributes? (All ratings were on a 10-point scale, with 1 being very low agreement and 10 being complete agreement with the statement.)

A stepwise multiple regression analysis was performed on the data to determine which of these attributes can be used to predict a person's likelihood of saving; and what the contribution of each attribute is to a person's likelihood of saving. The output is given in Exhibit 5.1.

EXHIBIT 5.1

Regression analysis (stepwise), annotated output

Correlation Matrix

	CONCERN 1	SAVE10 2	EARLY 3	RISK 4	NOSAVE 5	INS 6	SAVER 7	ADEQ 8	TOP 9	LIKELY 10
1. Concerned about rising $ of living	1.00	0.32	0.36	0.07	0.05	0.13	0.28	0.19	0.28	0.29
2. Should save 10% of income	0.32	1.00	0.52	0.06	-0.12	0.08	0.51	0.17	0.31	0.33
3. Establish retire. Plan early	0.36	0.52	1.00	0.14	-0.21	0.12	0.48	0.19	0.43	0.52
4. Accept risk for better rate	0.07	0.06	1.38	1.00	0.07	-0.03	0.04	0.08	0.13	0.12
5. Cannot afford to save	0.05	-0.12	-0.21	-0.07	1.00	0.06	-0.17	-0.12	-0.15	-0.28
6. Believe in insurance	0.13	0.08	0.12	-0.03	0.13	1.00	0.17	0.13	0.12	0.10
7. Put money away regularly	0.28	0.51	0.48	0.04	-0.17	0.17	1.00	0.39	0.38	0.43
8. Save adequately for retirement	0.19	0.17	0.19	0.08	-0.12	0.13	0.39	1.00	0.14	0.20
9. Saving for retirement top priority	0.28	0.31	0.43	0.13	-0.15	0.12	0.38	0.14	1.00	0.88
10. Likelihood of saving	0.29	0.38	0.52	0.12	-0.28	0.10	0.43	0.20	0.88	1.00

Bivariate correlations of all variables

Regression Analysis – Step 1

```
Equation Number 1   Dependent Variable. LIKELY
Block Number    1. Method:  Stepwise   Criteria PIN  .0500  POUT
.1000

Variable(s) Entered on Step Number
1.  TOP
```

Multiple R	**.88062**	R Square Change		.7755
R Square	**.77548**	F Change		3426.3727
Standard Error	.37628	Signif F Change		.0000

The first model includes the variable that is most highly correlated with the dependent variable 'Likely', which is 'Topprior'. It has an R-squared of 0.775 i.e., the model explains 77.5% of the variance.

```
Analysis of Variance
                    DF      Sum of Square      Mean Square
Regression           1        485.11920        485.11920
Residual           992        140.45123          .14158


F =    3426.37270     Signif F = .0000
AIC         .22542
CP       235.80466
SBC    -1931.33219


Equation Number 1            Dependent Variable.. LIKELY
```

─────────────────────Variables in the Equation─────────────────────

Variable	B	SE B	95% Confidence	Intrvl B	Beta
TOPPRIOR	**1. 531774**	.026168	1. 480422	1. 583125	.880615
(Constant)	**−2. 322437**	.101366	−2. 521353	−2. 123521	

The initial model with one variable is Likely = -2.53 + 1.32 (Topprior). The table also gives standard errors and the 95% confidence interval for the estimates.

─────────────────────Variables in the Equation─────────────────────

Variable	SE Beta	Correl	Part Cor	Partial	Tolerance	VIF
TOPPRIOR	.015044	.880615	.880615	.880615	1.000000	1.000

────── in ──────

Variable	T	Sig T
TOPPRIOR	58.535	.0000
(Constant)	−22.911	

─────────────────────Variables not in the Equation─────────────────────

Variable	Beta In	**Partial**	Tolerance	VIF	Min Toler	T	Sig T
CONCERN	.046483	.094237	.922784	1.084	.922784	2.980	.0030
SAVE10%	.123359	.247910	.906763	1.103	.906763	8.056	.0000
EARLY	.174114	.331128	.812035	1.231	.812035	11.047	.0000
RISK	.011854	.024817	.984093	1.016	.984093	.781	.4347
NOSAVE	−.154198	−.321927	.978605	1.022	.978605	−10.704	.0000
INS	−.873700	−.001831	.986370	1.014	.986370	−.058	.9540

| REGULAR | .111680 | .218260 | .857518 | 1.166 | .857518 | 7.041 | .0000 |
| ADEQUATE | .074143 | .154899 | .979964 | 1.020 | .979964 | 4.936 | .0000 |

Partial correlations (obtained after the entering the first variable) identify Early as the candidate to be added to the next model since this has the highest absolute partial correlation.

Regression Analysis – Step 2
Variable(s) Entered on Step Number 2
1. **EARLY**

Multiple R	**.89448**	R Square Change	.02462
R Square	**.80010**	F Change	122.03985
Standard Error	.35523	Signif F Change	.0000

Analysis of Variance

	DF	Sum of Square	Mean Square
Regression	2	500.51905	250.25952
Residual	991	125.05138	.12619

F =	1983.24239	Signif F = .0000
AIC	−2054.57448	
PC	.20111	
CP	103.4065	
SBC	−2039.86926	

————————————Variables in the Equation————————————

Variable	B	SEB	95% Confdence	Intrvl B	Beta
TOPPRIOR	1 **.400468**	**.027415**	1 **.346671**	1 **.454268**	**.805128**
EARLY	**.138561**	**.012543**	**.113947**	**.163174**	**.174114**
(Constant)	−2 .412973	.096046	−2 .601449	−2 .224495	

The second model includes two variables: 'Topprior' and 'Early'. The second model with two variables now reads: Likely = −2.41 + 1.40 (Topprior) + 0.138 (Early).

————————————Variables in the Equation————————————

Variable	SE Beta	Correl	Part Cor	Partial	Tolerance	VIF
TOPPRIOR	.015761	.880615	.725525	.851331	.812035	1.231
EARLY	.015761	.523176	.156899	.331128	.812035	1.231

——— in ———

Variable	T	Sig T
TOPPRIOR	58.535	.0000
(Constant)	−22.911	

——————— Variables not in the Equation ———————

Variable	Beta In	Partial	Tolerance	VIF	Min Toler	T	Sig T
CONCERN	.001892	.003910	.853567	1.172	.751125	.123	.9021
SAVE10%	.061373	.116568	.721135	1.387	.645799	3.693	.0002
RISK	−.002855	−.006306	.975611	1.025	.805036	−.198	.8427
NOSAVE	−.131537	−.287016	.951760	1.051	.789760	−9.427	.0000
INS	−.012628	−.027975	.980973	1.019	.807592	−.881	.3788

REGULAR	.055164	.105648	.733195	1.364	.694306	3.343	.0009
ADEQUATE	.052611	.115296	.960037	1.042	.795523	3.652	.0003

The process adding a variable is repeated (not shown here) until prespecified criteria are met.

Model summary

Step	MultR	Rsq	F (Eqn)	SigF		Variable	BetaIn
1	.8806	.7755	3426.373	.000	In:	TOPPRIOR	.8806
2	.8945	.8001	1983.242	.000	In:	EARLY	.1741
3	.9036	.8166	1469.028	.000	In:	NOSAVE	−.1315
4	.9051	.8192	1120.192	.000	In:	SAVE10%	.0603
5	.9059	.8206	903.946	.000	In:	ADEQUATE	.0388

The final table shows the final model: 5 variables – Topprior, Early, Nosave, Save10% and Adequate – are entered. The final R-squared is 0.8206 (82% of the variance explained).

5.4   Regression analysis: marketing applications

The regression applications here cover three different areas of marketing: product testing, demand forecasting and the importance of corporate brand loyalty in deciding to revisit a website.

What drives intention to revisit a website?

Marketing problem

The Internet has been gaining increasing acceptance as a marketing and distribution channel. Companies have been designing attractive websites to attract customers, while market researchers are busy studying the determinants of attitudes towards websites and reasons for visiting them. What is the importance of the corporate brand of well-known companies in generating revisits to their websites?

Magne Supphellen and Herbjørn Nysveen (2001) hypothesized that consumers loyal[1] to a brand will be motivated to evaluate all new information about the brand in a positive manner. By inference, they would attach their positive impression of the brand to the website instead of evaluating it independently of the brand. Any positive experience with the website will be enhanced and treated as a confirmation of the positive image already held, but negative experiences will be detached from the overall image and will be treated as isolated episodes. Consumers not loyal to the brand, on the other hand, will not bring in a positive bias to the evaluation. They will be more objective and their evaluation will be based on the performance features of the website. Consequently, their evaluation is likely to be less positive. This line of reasoning led the Supphellen and Nysveen to formulate three hypotheses. Consumers who are loyal to a corporate brand will:

1. evaluate the website of the corporation more positively;
2. probably evaluate the specific attributes of the website more positively;
3. intend to revisit the website.

[1] In this context, all reference to loyalty is to 'affective loyalty'.

How true are the generalizations implied in these hypotheses? This is the marketing problem tackled by this study.

Application of regression analysis

To understand how consumers evaluate websites, Supphellen and Nysveen interviewed 198 airline passengers who had visited the Scandinavian Airline Systems (SAS) website. Passengers at Oslo's Gardermøen Airport were approached and asked to fill in a questionnaire while waiting for their flights. Only those who had visited the SAS website were accepted for the final study.

The respondents rated the SAS website on a number of attributes. These were subjected to factor analysis and were reduced to three basic items: *functional attributes, safety* and *layout*. In addition, the respondents also rated on a seven-point scale their overall evaluation of the site (7 = very good; 1 = very bad):

- What is your overall evaluation of the site?
- How would you rate the overall quality of the site?

their loyalty (7 = strongly agree; 1 = strongly disagree)

- I feel that I have a relationship to SAS
- I prefer SAS even when other airline companies have better offers

and their intention to revisit (7 – strongly agree; 1 – strongly disagree)

- It is very likely that I will visit this website again.

Supphellen and Nysveen used overall evaluation of the site as the dependent variable and performed a series of hierarchical regressions on the data.

Results of regression analysis

Exhibit 5.2 shows the means, standard deviations and correlations among the attributes included in the regression analysis. Exhibit 5.3 shows the standardized regression coefficients for three models, all of which had overall attitude as the dependent variable:

EXHIBIT 5.2

Means, standard deviations, and correlations between study variables

	Attitude (site)	Intention to revisit	Functional attributes	Safety	Layout	Brand loyalty
Mean	4.4	5.9	4.3	4.9	4.2	3.6
Standard deviation	1.3	1.5	1.2	1,5	1.4	1.6
Correlations						
Attitude (site)	1.00					
Intention to revisit	0.34	1.00				
Function attributes	0.79	0.23	1.00			
Safety	0.12	0.13	0.12	1.00		
Layout	0.40	0.16	0.38	0.30	1.00	
Brand loyalty	0.30	0.27	0.23	0.13	0.18	1.00

EXHIBIT 5.3

Standardized regression coefficients for attitudes towards the website

Model	1	2	3
Dependent variable	Overall attitude	Overall attitude	Overall attitude
Functional attributes	**0.79**	**0.74**	**0.75**
Safety	−0.04	−0.03	−0.03
Layout	0.03	0.01	0.02
Brand loyalty		**0.15**	**0.14**
Functional × loyalty			0.01
Safety × loyalty			−0.03
Layout × loyalty			−0.05
Adjusted R^2 (as %)	62%	61%	60%

All statistically significant coefficients (at 0.05 level or better) are given in bold

Model 1. Site attributes only.
Model 2. Site attributes + brand loyalty.
Model 3. Site attributes + brand loyalty + interactions between attributes and brand loyalty.

Model 1 clearly shows that functional attributes of the website predict very strongly the overall attitude. Functional attributes feature very prominently in all models. If we add brand loyalty to this list, will it turn out to be a significant contributor to overall evaluation? Models 2 and 3 indicate that brand loyalty does indeed positively affect overall evaluation. However, consumers do not seem to evaluate the specific attributes of the website more positively because of their brand loyalty since the interaction effects (model 3) are not statistically significant.

Exhibit 5.4 shows similar results with intention to revisit as the dependent variable. Although the R^2 values are statistically significant, they are low. This can be attributed to many factors, such as the need to visit the site. What is of interest here is that brand loyalty is as strong a predictor as attitudes towards the site in predicting intent to visit.

EXHIBIT 5.4

Standardized regression coefficients for intention to revisit website

Model	1	2	3
Attitude towards the site	**0.34**	**0.23**	**0.21**
Brand loyalty		**0.21**	**0.21**
Attitude × loyalty			−0.12
Adjusted R-squared (as %)	11%	12%	13%

All bold values are significant at the 0.01 level.

How regression analysis addressed the problem
The results of regression analyses showed that brand loyalty is a significant factor in evaluating a website positively. Brand loyalty is also related to intention to revisit the website. The study also showed that individual features are not evaluated favourably merely because of brand loyalty. Supphellen and Nysveen believe that the study has the following implications to website designers and market researchers:

1. Website designers should make the corporate brand names very salient and vivid on their sites.
2. Brand managers need to understand the 'unique possibilities' of the web to highlight specific image elements or to build brand awareness.
3. Market researchers should be included in future studies of websites.

How to forecast the number of subscribers

Marketing problem
Forecasting sales is one of the most important tasks that a marketing manager can undertake. Regression-based forecasting models are often used in this context. In this illustration, the objective was to predict the number of cable TV subscribers for the years 1991–93, using a number of variables such as subscription rate, price index, adjusted monthly rate, disposable income, and number of households. Time series data were available for the years 1975–90. The analysis was carried out by Rao and Steckel (1995).

Application of regression analysis
The number of cable subscribers is closely related to the population size. So Rao and Steckel chose the number of cable TV subscribers per 100 000 as their dependent variable. Investigators hypothesized that two variables – *managerial decisions* as reflected in the monthly subscription rate for cable television, and *household economic status* as reflected in disposable income per household – would be useful predictors. Accordingly they used the following predictors: monthly subscription rate adjusted for inflation, and household disposable income as measured in constant 1987 dollars.

Results of regression analysis
Exhibit 5.5 shows the regression analysis results. The adjusted R-squared for this regression was 0.93, which implied a strong relationship between the predictor variables and the number of subscribers. The predictive equation was of the form

No. of subscribers = −70.69 + 0.0099 (Disposable income) − 2.10 (Monthly rate).

EXHIBIT 5.5

Regression coefficients

	Regression coefficient
Monthly rate	−2.10
Disposable income	0.0099
Intercept	−70.69
Adjusted *R*-square	**93%**

All coefficients were significant at the 0.05 level or better.

Both regression coefficients (monthly rate −2.10 and disposable income 0.0099) were statistically significant at the 0.10 level. (Although in scientific disciplines the commonly accepted levels of significance are 0.05 and 0.01, in marketing research 0.10 and 0.05 are more common.) The equation suggested that one unit of decrease in the monthly rate would increase the penetration rate of CATV by 2.1 units, while a $100 increase in disposable income would increase the penetration rate by 0.99 units.

How regression analysis addressed the problem

Since regression analysis provided the predictive equation which was found to be robust, predicting the subscriber base for 1991, 1992 and 1993 was a fairly straight-forward task: just plugging in the values for disposable income and monthly rate (in constant dollars) for the years for which predictions need to be made. This is shown in Exhibit 5.6. The results show that the actual results for these years were very close to the predicted figures. All prediction errors were positive for these three years (i.e., the model underpredicted the actual subscriber growth.)

EXHIBIT 5.6

Predicted and actual penetration rates

Year	Penetration		Error
	Actual	Predicted	
1991	55.77	51.91	6.9%
1992	56.76	54.94	3.2%
1993	58.41	57.87	0.9%

Models like these can also be used to set up what-if scenarios: What would happen if there were a huge tax cut and consumers were left with more disposable income? What would happen if we increased the cable subscription rate by 10%? What would happen if disposable income increased by a hundred dollars *and* we increased the subscription rate by 8%?

How relevant and consistent are the variables used in product testing?

Marketing problem

A major snack food manufacturer in the UK wanted to know whether the attributes consumers use to evaluate a confectionery brand were the right ones and whether these attributes remained consistent over time. This information would be helpful to the company in its product testing procedures. Leslie de Chernatony and Simon Knox (1990) examined the problem and designed an appropriate analysis to solve it.

Application of regression analysis

To address these issues, de Chernatony and Knox carried out two studies. In the first, a quota sample of 440 consumers in England, Scotland and Wales completed a monadic blind product test.[2] Two years later, another similarly selected and matched sample of 440 consumers completed a similar monadic blind test. The questionnaire was identical in both cases, with one exception: the 'propensity to buy' question was

---

[2]A *monadic* test is where the respondent rates a single product without comparing it to any other product. A *blind* test is one where the respondent does not know the identity of the product in terms of brand.

not included in the second survey. The questionnaire covered the following topics:

- Overall assessment of the brand after eating it (seven-point scale)
- Assessment of the brand on eight product attributes (three-point scale)
  - Taste of chocolate
  - Overall taste
  - Taste of centre filling
  - Overall smoothness
  - Light whipped centre
  - Overall softness
  - Softness of centre
  - Overall sweetness
  - Propensity to buy the brand (five-point scale)

If the eight product attributes are relevant and consistent over time (1) the same variables should be able to predict the respondents' overall assessment for both surveys (for testing purposes de Chernatony and Knox specified that these eight attributes should explain at least 70% of the variance); and (2) the relative importance of these eight attributes should be the same in both studies.

EXHIBIT 5.7

Varimax rotated component loadings year 1 vs. year 2

	Variance explained	
	Year 1	Year 2
Component 1*	26.7%	27.3%
Taste of chocolate	0.84	0.84
Overall taste	0.76	0.78
Taste of centre filling	0.79	0.73
Component 2*	20.3	22.2%
Overall smoothness	0.79	0.84
Light whipped centre	0.82	0.83
Component 3*	18.9%	21.6%
Overall softness	0.66	0.85
Softness of centre	0.76	0.82
Component 4*	12.8%	13.4%
Overall sweetness	0.94	0.93
Total**	78.7%	84.5%

* *Variance explained by the component.*
** *Variance explained by all four components.*

To get rid of the multicollinearity problem (see Section 5.5), de Chernatony and Knox applied principal components analysis with varimax rotation to each of the two sets of data and retained the components that were considered meaningful. This yielded four comparable components (Exhibit 5.7):

- Taste
- Smoothness
- Softness
- Sweetness

For each consumer, a factor score was computed for each of the four factors. In computing the component scores, de Chernatony and Knox used only those variables whose loading on a given factor exceeded 0.70. The amount of variance explained by each of these components also remained comparable (Exhibit 5.7) in both studies.

  Regression analysis was applied to these four component scores (independent variables) with 'overall assessment' as the dependent variable. This analysis was performed separately for each study.

Results of regression analysis
The results of regression analysis are shown in Exhibit 5.8. Did the variables have predictive ability? Only to a limited extent, since in either year these variables could not explain 70% of the variance as specified by the investigators. They only explained less than 60% of the variance. Were the variables consistent in their ability to predict? Again the answer is negative, since only taste was a consistent predictor in both years. Other variables were inconsistent. For instance, sweetness was a significant predictor in year 1 but not in year 3.

EXHIBIT 5.8
_____

Regression coefficients: Year 1 vs. Year 2

| | Regression coefficients | |
Attribute	Year 1	Year 2
Taste	0.85	0.85
Texture	0.27	0.12
Softness	0.28	0.16
Sweetness	0.09	0.18
Constant	0.36	0.81
R-squared (as %)	56%	59%

How regression analysis addressed the problem
Although the results turned out to be disappointing, they could be used to improve the process of identification and measurement of predictor variables. More specifically:

- The organization was not focusing on key intrinsic cues (variables that will have a strong impact on the overall assessment).
- Because branding is a key variable for this product, blind tests may not have been very effective.
- Lack of an open-ended question to assess what consumers liked/disliked may have limited the value of information collected.
- The scales used were inconsistent (seven-point for the dependent variable and three-point for independent variables) and were never tested for consistency or validity.

On the basis of this study, de Chernatony and Knox came to the following conclusions:

- In markets where branding is a strong evaluative cue, testing should include brand identification.
- Comparative (against the most popular brand in consumers' evoked set of brands) rather than monadic tests may be more appropriate in this context.
- The product being tested is probably a low-involvement category. Procedures that encourage consumers to consider their feelings about the product at a general level might have yielded better results.
- Consumers might have assigned scale values for the descriptive phrases used. These implied assignments might have been different for different consumers.

5.5   Caveats and concluding comments

While regression analysis is easy to perform and easy to analyse, a host of problems can affect the results. Careless and casual use of the technique can result in conclusions that are incorrect and misleading. Fortunately, most such problems are well documented and computer programs provide a number of diagnostics that would alert the researcher to their existence.

Collinearity

Collinearity occurs when two or more independent variables are highly correlated among themselves. In these situations the problem of confounding occurs because of the high intercorrelations: isolating the effect of correlated variables through partialling (which works very well when the correlations are not very high) does not work with highly correlated variables. The problem is known as *multicollinearity* or simply *collinearity*. Multicollinearity increases the variance of the regression coefficient, thereby making it unstable.

This problem is not always easy to identify by examining the bivariate correlation tables. However, most computer programs provide diagnostics to identify collinearity. This is generally done by calculating the squared multiple correlation (R^2) between each variable and all of the others. Some programs compute the *tolerance*, which is simply $1-R^2$. Another variation of R^2 is known as the *variance inflation factor (VIF)*, which is the reciprocal of tolerance: $1/(1-R^2)$. The researcher may choose to discard any variable based on a specified criterion depending on either of these measures (e.g., tolerance less than 0.10 or R^2 greater than 0.90).

When we encounter multicollinearity, we can combine the variables that are collinear; use alternative techniques such as ridge regression; or apply principal components analysis to the original data and substitute component scores for the values of measured variables.

Influential observations

Sometimes we get some data points that are atypical, commonly known as 'outliers'. These outliers affect the dependent variable in a way that is very different from other data points. These atypical observations can be so influential that they can actually change the regression line substantially. For instance, suppose we plot the relationship between income and the amount of money spent in restaurants by 30 people who work in an office. The regression line may look like Exhibit 5.9(a).

EXHIBIT 5.9

Regression of amount spent eating out on annual income, before and after removing an influential observation

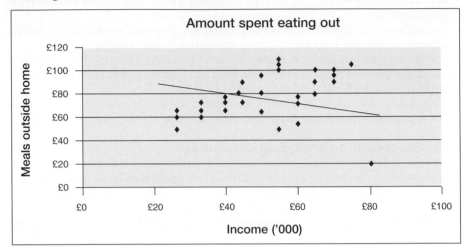

(a) Regression using all 30 data points

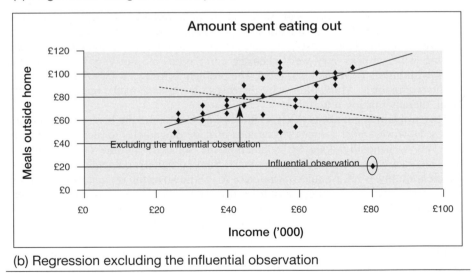

(b) Regression excluding the influential observation

However, if we look at the actual data points, we note that one high-income respondent spends hardly any money on outside meals. Removing this person from the analysis produces a different regression equation which fits the remaining respondents much better (Exhibit 5.9(b)). The data point removed from the analysis is known as the influential observation since this observation 'influences' the nature of the equation (for worse).

The most common reasons for influential observations are: (1) atypical behaviour of some individuals in a sample; and (2) mistakes in data entry. Since influential observations change the regression line such that it fits less well for the remaining observations, it is important to spot and eliminate influential observations. Influential observations exert much greater influence when the sample is small than when it is large. In marketing research the sample tends to be large, so this problem is not quite as common as other problems such as collinearity. However, there are cases in marketing research where it can be a major concern and therefore cannot be ignored.

In computer programs, outliers are identified by standardized residual scores (referred to as *distance*). Observations that deviate substantially from the mean (say, by more than three standard deviations) are examined, first, to make sure that these are not errors in data entry and, second, to see if these observations are influential, that is, whether removing them makes the regression equation different.

Another useful concept is *leverage*, or an observation's ability to affect the regression line. If we let h stand for leverage, p for the number of variables and n for the number of respondents, when observations are similar it will have mean of p/n. Any observation with an h greater than $2p/n$ should be examined carefully.

Influence can now be defined as

$$\text{Influence} = \text{Distance} \times \text{Leverage}.$$

So to assess whether an observation is influential, we need to consider both distance and leverage. The influence measure used by many computer programs is called *Cook's distance*. A distance of 1.0 or more indicates that the observation has an undue influence on the regression equation and remedial actions may have to be taken.

Extrapolation

Regression analysis is often carried out for predictive purposes such as sales forecasting. For example, by using regression analysis we can identify the relationship between sales and other variables (such as inflation rate, interest rate and price of the product) over the past several years. We can then use this equation to predict the sales in the future. This extrapolation of the equation to data outside the range can be risky. This is because the validity of the equation outside the range is not tested and cannot be taken for granted. Relationships that worked for the period 1980–2000 may not work very well if applied to 2000–2020. In general, extrapolation of the equation to points that are closer to the model range is less risky than those that are farther away. For instance, applying a forecasting equation developed on data for years 1990–2002 to the years 2003 and 2004 is likely to be less risky than applying the same equation to 2009 and 2010.

Non-linearity

Regression analysis assumes that the relationship between the dependent and the independent variables is a straight line. This may not always be the case. For instance, if we relate actual income to the percentage spent on luxury items, we may find that the more you earn, the more you spend on luxury goods. However, the overall relationship may not be linear. For instance, the highest-income group may spend less money on luxury goods (as a proportion of their income) than the middle-income group. Yet within each of these income groups, there may be a linear relationship between income and proportion of income spent on luxury goods. Non-linearity can be identified by looking at the residuals. If residuals exhibit a non-random pattern on either side of the mean, then the relationship may be non-linear.

Many solutions are available to deal with such problems. We can use regression methods that are specifically available for non-linear data. Or we can divide the observations into homogeneous segments and try regression again. A third solution is to apply some transformation to the data to make it linear, apply linear regression and retransform the data. The type of transformation required (log, $1/x$, etc.) will depend the nature of the data. (Most major statistical packages will be able to transform the data to the researcher's specification.)

Small samples in relation to the number of variables

If the sample size is not substantially larger than the number of variables, then it is possible to get large R^2 values even when there is no relationship. The expected value of R^2 is given by:

$$E[R^2] = \frac{p}{n-1}$$

where $E[R^2]$ is the expected value of R^2, p is the number of variables, and n is the sample size. Thus if we perform a regression on 20 variables on a base of 41 respondents, we can expect to find an R^2 of 0.5 purely by chance. To avoid this problem it may be a good idea to keep the number of respondents 'large' relative to the number of variables. A rule of thumb would be to have at least 10 respondents per variable, so that in the above example we should aim to have a minimum of 200 respondents. It is true that in many cases this is not possible. For instance, in forecasting sales using 10 variables, we may not have 100 years (or even 100 quarters) of data. Even if we had, the relationship of these variables to sales may have changed over the years. In such cases, R^2 values that do not exceed the expected value as computed above should be used with extreme caution.

Identifying dependent and independent variables

Consider a regression equation in which the investigator tries to predict overall customer satisfaction through a number of variables such as customer evaluation of the company's product quality, service quality, price, etc. On the face of it, the model looks reasonable. But is the customer satisfied overall because of price acceptance or did overall satisfaction with the company lead to price acceptance? Marketing actions will depend on the nature of such relationships. If satisfaction leads to price acceptance, then the marketer would attempt to increase customer satisfaction. If price acceptance led to satisfaction, then the marketer might find ways to ensure that the price is acceptable. Identifying the nature of relationships is not a statistical problem. So the researcher should pay a lot of attention to the reasonableness of the model as specified.

The conceptual simplicity of regression analysis, and the easy availability of computer packages to implement it, hide a myriad of traps that an inexperienced researcher may fall into. Regression analysis, one of the most useful of all statistical techniques, also demands considerable theoretical expertise and interpretive skills.

Bibliography

Further reading

For a detailed exposition of regression analysis readers are referred to

Draper, Norman R. and Smith, Harry (1998) *Applied Regression Analysis* (3rd edition). Wiley, New York.

For a leisurely, practical and less technical exposition, written in a question-and-answer format, see:

Allison, Paul D. (1999) *Multiple Regression*. Pine Forge Press, Thousand Oaks, CA.

References

de Chernatony, Leslie and Knox, Simon (1990) How an appreciation of consumer behaviour can help in product testing. *Journal of the Market Research Society*, **32**(3), 329–48.

Rao, Vithala and Steckel, Joel H. (1995) *The New Science of Marketing*. Irwin, Chicago.

Supphellen, Magne and Nysveen, Herbjørn (2001) Drivers of intention to revisit websites of well-known companies. *International Journal of Market Research*, **43**(3), 341–52.

6

Discriminant Analysis

6.1   What is discriminant analysis?

A fundamental theme in marketing is identifying how a consumer would behave. Will the customer be a good credit risk or a bad one? Will the consumer buy your product or that of your competitor? Will the price increase be acceptable to consumers or not? These questions involve classifying respondents into different groups. This is usually accomplished by identifying the importance of variables that are related to an individual's belonging to a group and creating a linear combination of these variables that best predicts group membership.

Consider a bank trying to differentiate customers as good or bad credit risks on the basis of their income and net worth. Exhibit 6.1 shows the distribution of customers

EXHIBIT 6.1

Creditworthiness vs. income and net worth

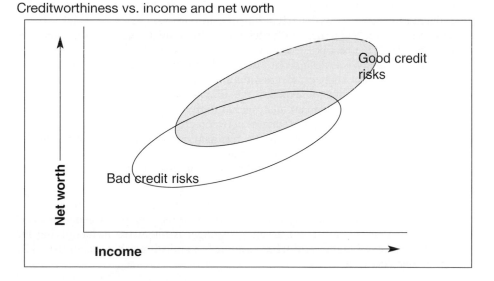

EXHIBIT 6.2

Eliminating bad credit risks

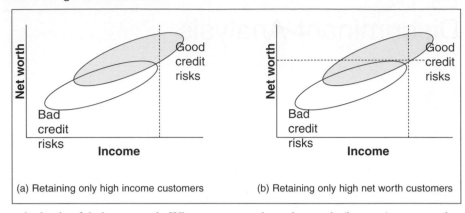

(a) Retaining only high income customers   (b) Retaining only high net worth customers

on the basis of their net worth. When we move along the *x*-axis (income), we note that as the income increases, so does creditworthiness (good credit risks). Similarly, on the *y*-axis (net worth), as net worth increases, so does creditworthiness.

If the bank confined itself to customers at the highest income level, it might minimize its bad credit risk customers. However, by doing so, it will miss a large number of good credit risk customers whose income is relatively low. We can do similar analysis with net worth going along the *y*-axis. Here again, we have a similar pattern. The bank can avoid bad credit risks by catering only to high net worth customers, but by doing so it will miss a huge number of profitable customers whose net worth may not be high (Exhibit 6.2). So the bank has this problem: how to include as many good credit risk customers as possible while at the same time excluding as many bad credit risk customers as possible.

One way to accomplish this is to create a linear combination of the existing variables (net worth and income). This is the equivalent of creating a new variable which is the weighted combination of the original variables such that the two groups are differentiated as clearly as possible (Exhibit 6.3). In our example, what weight should we assign to income and to net worth such that it differentiates good credit risks from bad credit risks in the best possible way?

Discriminant analysis is designed to solve problems like these. It is best understood in terms of regression analysis. Regression analysis assigns a weight to each independent variable such that it explains the maximum amount of variance in the dependent variable. Discriminant analysis similarly assigns a weight to each independent variable such that it differentiates between groups. In regression analysis the dependent variable is metric, while in discriminant analysis it is non-metric. In this way, both techniques are conceptually similar.

Discriminant analysis is used in several contexts in marketing. Here are some typical applications:

1. A large bank has extensive information on its customers such as their age, gender, income, average monthly balance, and credit records, profit generated, and credit payment history. The marketing manager is interested in identifying the characteristics of highly profitable customers and distinguishing them from less profitable customers.

EXHIBIT 6.3

Creditworthiness as a function of income and net worth

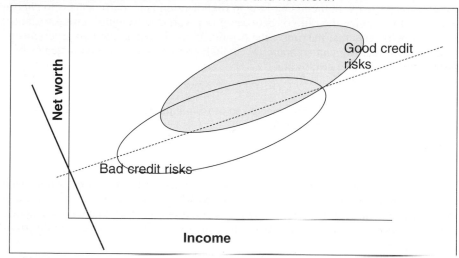

2. A computer manufacturer has consumer evaluations of different brands on several attributes. Based on these evaluations, can we assess how customers distinguish between two different brands? What attributes contribute to product distinction and to what extent?
3. A marketing manager of a luxury car would like to identify the attributes that are effective in distinguishing buyers and non-buyers of luxury cars.

In all these cases the aim is to identify the variables that distinguish the groups, and assign weights to each variable so that the distinction is maximized. Once we develop such weights, we can also use the linear combination to assign future observations to groups. For instance, if a bank identifies the linear combination of attributes that best distinguishes good and bad credit risks, it can then apply this equation to the attributes of a potential customer to identify whether he is a good risk or not.

6.2   The discriminant analysis model

We have a number of independent metric variables (such as customer ratings of different product attributes) and a single non-metric dependent variable (such as whether the consumer is a buyer or a non-buyer of a brand). How are a consumer's ratings on product attributes related to buying or non-buying? To answer this question we can use discriminant analysis and weights for each of the independent variables that would best predict whether the customer is a buyer or a non-buyer. We start our discussion with a two-group discriminant analysis (i.e., a dependent variable with two groups).

The discriminant function

If we let Z stand for the discriminant function that best differentiates the two groups – buyers and non-buyers – then

$$Z = a_1 x_1 + a_2 x_2 + \dots + a_p x_p,$$

where $x_1, x_2, x_3, \dots, x_p$ are values of independent variables and $a_1, a_2, a_3, a_4 \dots a_p$ are weights associated with them. One way to find discriminant coefficients is to choose values that maximize the ratio of between-group variance to within-group variance, i.e., the coefficients that maximize the F ratio. If we designate the two group means as $\overline{Z_1}$, and $\overline{Z_2}$, then we can compute D^2 (Mahalanobis distance) which indicates how farther apart these two groups are:

$$D^2 = (\overline{Z_1} - \overline{Z_2})^2 / S_z^2$$

The values of $a_1, a_2, \dots, a_p$ are selected so as to maximize D^2, so that the groups are as clearly distinguished as possible.

Cutoff score

The value of Z in the discriminant equation is similar to the value of y' in the regression equation. To use this value to group respondents we need a cutoff score. Each respondent is assigned to a group depending on whether the respondent's Z score is above or below this cutoff score. The cutoff score is chosen such that it *minimizes the number of misclassifications*. To obtain the cutoff score (C) for two-group discriminant analysis:

1. Apply the discriminant function to the means of each group separately to obtain $\overline{Z_1}$ and $\overline{Z_2}$.
2. If both groups have the same sample size, $C = (\overline{Z_1} + \overline{Z_2})/2$. If the groups have different sample sizes, $C = (n_1\overline{Z_1} + n_2\overline{Z_2})/(n_1 + n_2)$.

There is a more complicated quadratic rule for assigning respondents to groups. However, Huberty (1984) points out that results obtained by a linear rule yield greater stability than those by a quadratic rule when small samples are used and when the normality conditions are not met.

Standardized coefficients

As with the regression coefficients, the values of $a_1, a_2, \dots, a_p$ are not directly comparable. We can obtain the relative importance of the discriminant coefficients by standardizing them. Standardized discriminant coefficients can be obtained by multiplying the a_i by the corresponding pooled standard deviations. From the size of standardized discriminant coefficients we can infer the ordinal importance of variables $x_1, x_2, \dots, x_p$.

Probability of group membership

Although so far we have assigned respondents to one or the other group, there is a chance of misclassification. To assess the chance of misclassification, we can compute the actual probability of a respondent belonging to a group:

$$\text{Probability of belonging to group } 1 = \frac{1}{1 + \exp(-Z + C)}.$$

For instance, if respondent A has a probability of 0.51 of belonging to group 1 and respondent B has a probability of 0.87, it would mean that B is a much stronger candidate for group membership than A.

In most computer programs, these probabilities are referred to as posterior probabilities. They can be valuable to marketers in that they may be more interested in consumers who have high probability of belonging to a group (such as those who have a high propensity to buy a brand) than those who only have marginal tendency to do so.

How good is the discriminant function?

To validate the discriminant function obtained, we can apply it to each respondent and predict his or her group membership. We can then compare this with their actual group membership. Most computer programs will provide a table something like this:

Actual group	Predicted group	
	1	2
1	*a*	*b*
2	*c*	*d*

Cell *a* shows the number of respondents whom the discriminant function correctly predicted to be members of group 1; cell *d* shows the number of respondents whom the discriminant correctly predicted to be members of group 2; cells *b* and *c* show respondents who were misclassified. This procedure tends to inflate the validity of the procedure since we have used the same sample to develop the equation *and* to validate it. Ideally, we would like to derive the equation from one set of data and apply it to another set of data.

One way to validate the results is through using what is known as a *holdout sample*. Suppose we have a sample of 1000 respondents; we use 500 to develop the equation and use the remainder (the holdout sample) to validate the equation. This is a commonly used procedure.

If the sample is small, the holdout sample may not be a good procedure to use, since splitting the sample further reduces the sample size and can potentially make the estimates less stable. In such cases, one can use the *jackknife procedure*. Here we exclude one observation at a time from the sample, estimate the equation, then apply the equation to the excluded observation. We repeat this for all respondents. There are computer programs available which execute such procedures.

Selecting the best set of predictors

The procedures given for regression analysis – hierarchical and non-hierarchical methods – apply here as well. Whereas in regression analysis we tested to see whether R^2 was altered when we added or deleted a predictor, in discriminant analysis we test whether the value of D^2 is altered by deleting or adding a variable.

The appropriate test statistic is

$$F = \frac{(n_1 + n_2 - p - 2)(n_1 n_2)(D_{p+1}^2 - D_p^2)}{(n_1 + n_2)(n_1 + n_2 - 2) + n_1 n_2 D_p^2)}$$

with 1 and $n_1 + n_2 - p - 2$ degrees of freedom, where p is the number of variables in the equation. The critical value is the tail area to the right of the computed statistic.

As with regression analysis the user may specify a criterion for adding or deleting a variable (such as F-to-enter or F-to-remove) in computer programs using hierarchical methods.

The overall significance of the equation can be tested with the null hypotheses that none of the variables improve classification based on chance alone. This can be tested by the F statistic

$$F = \frac{n_1 + n_2 - p - 1}{p\,(n_1 + n_2 - 2)} \times \frac{n_1 n_2}{n_1 + n_2} \times D^2$$

with p and $(n_1 + n_2 - p - 1)$ degrees of freedom.

Using categorical independent variables

Discriminant analysis accommodates categorical independent variables. They are specified in the same way as in regression analysis (see Chapter 5 for details).

Discriminating among more than two groups

Although conceptually multi-group discriminant analysis is an extension of two-group discriminant analysis, both in terms of mathematics and general interpretation multi-group discriminant analysis can be complex. Here we provide only a very brief overview of multi-group discriminant analysis.

In multiple discriminant analysis, classification functions are computed for each group. The approximate F statistic tests the null hypothesis that the means of all groups are equal for all variables simultaneously. F statistics are also computed for pairs of groups to test the equality of means of the two groups. The analyst may use this statistic to decide whether to combine some groups if they are fairly close.

To classify an individual into one of the groups, we evaluate an individual's scores on different variables using discriminant functions corresponding to each group. The individual is assigned to the group for which the computed classification function is the highest.

Prior probabilities

Consider a case in which the probability of an individual falling in group A is 0.9 and in group B 0.1. In this case, we can achieve 90% accuracy by simply allocating everyone to group A. To avoid this ineffective solution, the allocation procedure that assigns an individual to a group based on Mahalanobis distance should be modified. Many computer programs allow the user to enter prior probabilities for this purpose.

Assumptions

Most methods discussed here are based on certain assumptions. The first assumption is that the within-group covariance matrix is the same for all groups. The second assumption is, for tests of significance, that the data should be normally distributed within groups. However, as Manly (1994) points out, even if both assumptions are violated we might still be able to discriminate among groups, although it may not be simple to establish the statistical significance of group differences.

6.3   Alternative technique: logistic regression

Logistic regression

The objectives of logistic regression are similar to those of two-group discriminant analysis. It is a good alternative to discriminant analysis when we cannot assume a multivariate normal distribution. Logistic regression analysis can be used for any combination of metric and non-metric variables. The dependent variable takes on a value of 1 or 0 (e.g., a respondent is a buyer or a non-buyer). What logistic regression attempts to do is predict the *probability* of a person belonging to a group. For example, logistic regression enables us to state (on the basis of the values of the independent variables) that the probability of a consumer buying our product is 0.8. However, probability (p) could not be directly used as a dependent variable for many reasons. Consider this: if we express p as the linear combination of independent variables and in an ordinary regression, the value of p could exceed 1.0. To avoid this problem, we make a logistic transformation of p (the logit of p): logit (p) is the natural log of the odds or likelihood ratio that the dependent variable is 1,

$$\text{logit } (p) = \log \left(\frac{p}{1-p} \right).$$

The value of logit (p) can range from negative infinity to positive infinity. The logit of 0.5 is 0, and the scale is symmetrical around this value, as shown in Exhibit 6.4. The logit scale is approximately linear in the middle range and logarithmic at extreme values: the difference in logits between ps of 0.95 and 0.99 is much bigger than that between 0.5 and 0.8. Logistic regression is of the form

$$\text{logit } (p) = b_0 + b_1 x_1 + b_2 x_2 + b_3 x_3 + \dots .$$

EXHIBIT 6.4

The relationship between probability (p) of group membership and the corresponding logit (p)

p	0.3	0.4	0.5	0.6	0.7	0.8	0.9	0.95	0.99
logit (p)	−0.847	−0.405	0.0	0.405	0.847	1.386	2.197	2.944	4.595

To enter the independent variables, we can use stepwise (forward or backward elimination) or simultaneous methods. However, unlike linear regression, which uses the least-squared deviations criterion for the best fit, logistic regression uses a maximum likelihood method that maximizes the probability of getting the observed results given the fitted regression coefficients. Consequently, the goodness of fit and overall significance statistics used in logistic regression are different from those used in linear regression. We can also use the percentage of correct classifications (as in discriminant analysis) table to assess the effectiveness of the model.

When we calculate the likelihood of observing the exact data we actually did observe under the null hypothesis (all coefficients are 0) and the alternative hypothesis (the model is correct), the resulting number tends to be small in magnitude. Since it is more convenient to handle larger numbers, we calculate its natural logarithm. This gives us the log-likelihood. Because probabilities are always less than 1, log-likelihoods are always negative. Again for convenience, we work with negative log-likelihoods.

Interpreting logistic regression coefficients

As with multiple regression, logistic regression holds that, when other variables in the equation are held constant, there is a change of b_i in logit (p) for every unit increase in x_i. Since the logit transformation is non-linear, the change in p associated with a unit change in x_i will vary with the value of x_i. However, there is a straightforward way to interpret a constant increase in the value of x_i. It involves a constant *multiplication* [by exp(b)] of the odds that the dependent variable takes the value 1 rather than

0. For example, if b_i takes the value 2.30, exp (2.30) equals 10. This means if x_i increases by 1, the odds that the dependent variable takes the value 1 increase tenfold. So, when x_i takes the value 0, logit (p) is 0; this means that there is an even chance of the dependent variable taking the value 1. If x_1 increases to 1, the odds that the dependent variable takes the value 1 rise by a factor of 10, from 1:1 to 10:1, to a p of 0.909. If x_i increases to 2, then the odds will be 100:1, a p-value of 0.990. We can represent the results of logistic regression through a plot that shows the odds change produced by unit changes in different independent variables.

A frequently used statistic to assess the statistical significance of each independent variable is the Wald statistic. The Wald statistic has a chi-squared distribution. It is interpreted in the same way as the t-values for each independent variable in multiple regression analysis.

The relative importance of the independent variables can be assessed by multiplying each coefficient by the standard deviation of the corresponding variable. The ranking of the resultant values will reflect relative importance of the independent variables.

6.4   Discriminant analysis: computer output

In the survey of non-savers mentioned in the previous chapter, the respondents were also asked if they had dependants. Given a person's response to the attributes listed in Section 5.3, can we predict whether he or she has dependants? A stepwise discriminant analysis was performed on the data. The output is given in Exhibit 6.5.

EXHIBIT 6.5

Discriminant analysis output, annotated

Discriminant – Group means and Standard deviations

```
Number of Cases by Group
            Number of Cases
SAVE        Unweighted      Weighted      Label
1             514             514.0       Have dependants
2             496             496.0       No, do not have dependants
Total        1010            1010.0
```

```
Univariate F-statistics
Wilks' Lambda (U-statistic) and univariate F-ratio
with 1 and 1008 degrees of freedom
```

Variable	Wilks' Lambda	F	Significance
CONCERN	.96727	34.10	.0000
SAVE 10%	.87179	148.20	.0000
EARLY	.78037	283.70	.0000
RISK	.99489	5.18	.0231
SAVE DIFFICULT	.86660	155.20	.0000
INS	.99929	.72	.3981
REGULAR	.86768	153.70	.0000
ADEQUATE	.99037	9.81	.0018

Wilk's lambda is used as the selection criterion. At each step, a variable is added (or deleted) from the discriminant function based on lambda.

```
On groups defined by SAVE
Analysis number 1

Stepwise variable selection

      Selection rule: Minimize Wilks' Lambda
      Maximum number of steps                    42
      Minimum Tolerance Level               .00100
      Minimum F to enter                    1.0000
      Maximum F to remove                   1.0000
```

A tolerance of 1.0 was used.

```
Canonial Discriminant Functions

      Maximum number of functions                  1
      Minimum cumulative percent of variance  100.00
      Maximum significance of Wilks' Lambda   1.0000
      Prior probability for each group is   .50000
```

The prior probability is set to be equal to 0.5.

```
At step 1, EARLY was included in the analysis.
```

Between Groups		Degrees of Freedom			Signif.
Wilks' Lambda	.78037	1	1	1008.0	
Equivalent F	83.692		1	1008.0	.0000

```
============= Variables in the analysis after step 1 =============
```

Variable	Tolerance	F to remove	Wilks' Lambda
EARLY	1.0000000	283.69	

The first variable entered is the one with the lowest Wilk's lambda, Early.

```
=========== Variables not in the analysis after step 1 ===========
Minimum
```

Variable	Tolerance	Tolerance	F to enter	Wilks' Lambda
CONCERN	.9014281	.9014281	.26330	.78017
SAVE 10%	.8170169	.8170169	23.573	.76252
RISK	.9860240	.9860249	.64001	.78032
SAVE DIFFICULT	.9976471	.9976471	105.87	.70613
INS	.9860387	.9860387	1.0362	.77957
REGULAR	.8572366	.8572366	33.112	.75553
ADEQUATE	.9736971	.9736971	.12793	.78027

The table above shows the changed values of Wilk's lambda after the entry of the first variable is entered in the model. The next variable to be entered will be the variable with the lowest lambda in the above table.

This process is continued (output not shown here) until prespecified criteria are met.

Summary Table

Step	Entered Removed	In	Lambda	Sig	Label
1	EARLY	1	.78037	.0000	Ret.Plans s/b established
2	SAVE DIFFICULT	2	.70613	.0000	It will be difficult to save
3	REGULAR	3	.68787	.0000	Should save regularly
4	SAVE 10%	4	.67992	.0000	Should save 10% of inc.
5	ADEQUATE	7	.67181	.0000	Expect to save for ret.
6	CONCERN	8	.66985	.0000	Concerned about rising cost of living

The summary table shows the variables retained in the final model.

```
Classification Function Coefficients
(Fisher's Linear Discriminant Functions)
```

Save =	1	2
CONCERN	3.496813	3.320905
SAVE10%	2.852239	2.453510
EARLY	2.737563	1.647793
SAVEDIFFICULT	1.581034	2.331002
REGULAR	.9590767	.3945166
ADEQUATE	−.3096838	−.6804376
(constant)	−46.96494	−40.97184

The weights of the two classification functions above are applied to each indivdual's ratings. An individual is then assigned to the group with the largest classification score. The discrimination function can be obtained by subtracting the value of classification function 2 from the corresponding value of classification function 1.

```
Discriminant Analysis
```

Fcn	Eigenvalue	Pct of Variance	Cum Pct	Cannoical Corr	After Fcn	Wilks' Lambda	Chisquare	DF	Sig
					: 0	.6682	404.129	11	
1*	.4965	100.00	100.00	.5760	:				

```
* marks the 1 canonical discriminant functions remaining in the
analysis
```

```
Standardized Canonical Discriminant Function Coefficients
```

	FUNC1
CONCERN	.07773
SAVE 10%	.18454
EARLY	.53970
SAVE DIFFICULT	−.54418
REGULAR	.27566
ADEQUATE	−.13597

As with regression, the relative effect of different variables can be obtained from the standardized canonical discriminant coefficients.

Structure Matrix: Pooled-within-groups correlations between
discriminating variables and canonical discriminant functions

(Variables ordered by size of correlation within function

	FUNC 1
EARLY	.75290
SAVE DIFFICULT	−.55683
REGULAR	.55420
SAVE 10%	.54425
CONCERN	.26105
ADEQUATE	.13997
INS	.04711

The structure matrix gives the loadings (correlations) between the original variables and the discriminant score.

Canonical Discriminant Functions evaluated at Group Means (Group
Centroids)

Group	FUNC 1
1	.34541
2	−1.43453

Discriminant Analysis

Classification of Respondents Using Discriminant Analysis

Actual	No. of Group Cases	Predicted Group Membership 1	2
Group 1	514	423 82.2%	91 17.8%
Group 2	496	101 20.4%	156 79.6%

Percent of 'grouped' cases correctly classified: 81.68%

The table is the result of applying the classification coefficients to the data. The figures in the main diagonal shows correct classifications. Overall 81.68% of respondents were correctly classified.

6.5 Discriminant analysis: marketing applications

Discriminant analysis is well suited for solving a variety of marketing problems.
Marketers are always interested in differentiating one type of customer from another:
What is the difference between those who buy brand A and those who buy brand B?
What influences a patient to choose a walk-in clinic over a family doctor? What differentiates a loyal customer from a non-loyal one? Here we look at in some detail the
application of discriminant analysis to a few real-life problems.

Why do some choose walk-in clinics over private physicians?

Marketing problem

In the United States there has been a trend towards the use of these clinics. The increase in these clinics has been fuelled by a need to reduce medical care costs. The proliferation of walk-in clinics has implications for insurance companies, Medicare, patients and private investors in those facilities. As a result, it is important to know whether there are differences between patients who go to walk-in clinics as opposed to private physicians. In particular, it will be of interest to know: what consumers expect of these two healthcare delivery systems; how their evaluations of them differ; and differences between the demographic characteristics of the two groups of patients. Dant, Lumpkin and Bush (1990) addressed this issue through a research project that utilized discriminant analysis. They felt that such information would enable healthcare providers to identify the markets better, offer the expected services and reduce costs by eliminating services that are not considered important.

Application of discriminant analysis

From prior literature and preliminary qualitative research, Dant *et al.* identified the following 10 attributes to be of importance in distinguishing these two groups of consumers:

- Convenient location
- Friendly staff
- Good reputation
- Different specialists in the same building
- Ability to get appointments quickly
- Short waiting time
- Reasonable costs
- Friendly doctors
- Competence of doctors
- Physician's willingness to spend time and explain.

They formulated the following hypotheses:

1. Users of walk-in clinics will have higher *expectation determinants* for attributes such as convenience of location, lower cost of services, ease of gaining appointments and flexible hours of operations compared to users of private physicians.
2. Users of walk-in clinics will have higher *performance evaluations* for attributes such as convenience of location, lower cost of services, ease of gaining appointments and flexible hours of operation compared to users of private physicians.
3. Need and treatment requirements of walk-in clinics centre on emergency care procedures, whereas for users of private physicians they centre on non-emergency procedures, physical examination and preventive health programmes.

In addition, Dant *et al.* were interested in understanding how demographics relate to expectations and performance evaluations.

A sample of 670 was randomly drawn from 15 different cities in five contiguous states in the southwestern USA. To qualify for the study, an adult should have personally visited (or been taken to) a medical doctor in the past six months. All interviews were conducted over the telephone and, of the 670 interviews, 602 were deemed to be complete enough to be included in the analysis. Of these, 397 were

users of private physicians and 205 of walk-in clinics.
  Respondents were asked to:

1. Rate the importance of the 10 characteristics listed above on a three-point scale: unimportant (1), somewhat important (2) and very important (3).
2. Rate the extent to which their provider (walk-in or private physician) had these 10 characteristics: not at all (1), only somewhat (2), very much so (3).
3. Indicate their demographic characteristics such as marital status, education, gender, income, etc.

In addition to other analyses, discriminant analysis was applied to the above data with a view to understanding the differences between patients of walk-in clinics and private physicians.

Results of discriminant analysis
The results of discriminant analysis for *expectations* are shown in Exhibit 6.6. This exhibit answers the question whether the expectations of walk-in clinic users are different from those of patients of private physicians. The means on each attribute for the two groups are very close, and only for two out of the ten attributes were the differences not statistically significant. The overall hit rate (correct classification of patients on the basis of the discriminant function) was 66%, whereas the proportional chance criterion was 55%. Dant *et al.* concluded that since the actual prediction was not even 25% higher than the chance criterion and only two of the 10 attributes were statistically significant, the expectations of the two groups could not be considered different.

  When Dant *et al.* applied discriminant analysis to *performance* (Exhibit 6.7), they did not fare much better. Seven out of ten attributes did not exhibit statistically

EXHIBIT 6.6

Discriminant analysis: importance of expected attributes

	Disc. loading	Group means		F ratio
		Clinics	Physicians	
Convenient location*	0.55	2.36	2.21	6.14
Friendly staff	−0.32	2.51	2.58	2.07
Good reputation	−0.42	2.80	2.86	3.56
Different specialists in same building	−0.04	1.82	1.83	0.03
Ability to get appointment quickly	0.37	2.66	2.58	2.72
Short waiting time*	0.54	2.60	2.47	5.75
Reasonable costs*	0.27	2.58	2.51	1.44
Friendly doctors	−0.09	2.68	2.70	0.14
Competence of doctors	−0.13	2.91	2.92	0.31
Physicians' willingness to spend time and explain	−0.07	2.86	2.87	0.08
Hit ratio (correct classification)				
Actual overall	66%			
Proportional chance criterion	55%			

* *Statistically significant differences*

EXHIBIT 6.7

Discriminant analysis: performance of attributes

	Disc. loading	Group means		F ratio
		Clinics	Physicians	
Convenient location*	0.52	2.43	2.29	6.03
Friendly staff	0.06	2.59	2.57	0.09
Good reputation*	−0.66	2.67	2.80	9.50
Different specialists in same building	−0.01	1.84	1.84	0.00
Ability to get appointment quickly	0.36	2.58	2.50	2.82
Short waiting time	0.23	2.43	2.37	1.13
Reasonable costs	0.16	2.42	2.38	0.58
Friendly doctors	−0.04	2.72	2.73	0.03
Competence of doctors*	−0.44	2.80	2.87	4.29
Physicians' willingness to spend time and explain	−0.18	2.74	2.77	0.69

Hit ratio (correct classification)
Actual overall	67%
Proportional chance criterion	55%

* *Statistically significant differences*

significant differences between the two groups. The overall hit rate of 67% fell short of the 'plus 25%' over the chance criterion (55%). Dant *et al.* concluded that the two groups rated the care they received similarly.

Stepwise discriminant analysis applied to demographic characteristics (Exhibit 6.8) resulted in a hit ratio of 72% which fulfilled the 'plus 25%' over the chance criterion (55%). Dant *et al.* concluded that those who visit clinics have a different demographic profile to those who go to private physicians.

How discriminant analysis addressed the problem

The study showed that there are differences in the criteria used for the selection of healthcare providers. However, the differences are not as pronounced as had been believed. While there are significant demographic differences among the two groups, this has not translated into attitudinal differences, either in terms of expectations or in terms of perceived performance characteristics.

While conventional wisdom might suggest two distinct market segments seeking two distinct modes of healthcare delivery, discriminant analysis indicates that the two groups may not be that distinct after all. They expect similar things, and perceive having received similar benefits from the mode of healthcare they use. The mode of care the two groups seek may have more to do with their demographic profiles than with what they expect or receive in terms of healthcare benefits.

Can you reposition milk as a soft drink?

Marketing problem

The soft-drink market in New Zealand is very competitive, with the market leaders spending up to $500 000 a year on television advertising alone. (Just to put this in perspective, New Zealand's population of 3.8 million is about 1½% the size of that of the

EXHIBIT 6.8

Discriminant analysis: Demographic characteristics

	Disc. Loading	Clinics	Physicians	F ratio
Marital status				
Divorced	0.00	29.09	70.91	0.72
Married	0.01	30.48	69.52	4.57
Widowed	0.00	17.24	82.76	0.37
Never married		48.20	51.80	
Education				
Some college	0.00	38.42	61.58	2.21
Graduate degree		29.82	70.18	
Sex of respondent				
Male	0.01	28.65	61.35	4.83
Female		30.06	69.94	
Income ($)				
20,000 or less	0.03	44.26	55.74	7.78
20,001–30,000	0.00	40.94	59.06	3.11
50,000 or more		27.00	73.00	
Number of children at home				
One	0.03	24.79	74.21	14.35
Two		36.36	63.64	
Three or more		39.02	60.98	
Age of youngest child at home				
10 or younger	0.03	35.80	64.20	44.27
11–20 years		26.97	73.03	
21 or older		20.00	80.00	
Type of medical need				
Major illness/injury	0.01	42.86	57.14	6.02
Routine/follow-up	0.03	25.83	74.17	24.13
Emergency	0.03	58.44	41.51	16.43
Other		37.70	62.30	
Hit ratio (correct classification)				
Actual overall	72%			
Proportional chance criterion	55%			

USA.) The market is characterized by heavy consumption among image-conscious teenagers.

The Dairy Board faced a major task in its attempt to penetrate this market with its flavoured milk, Zap. Major established brands such as Coca-Cola and Fanta immediately evoked mental images of fun, excitement and peer-group acceptance, while milk was regarded as nutritious, healthful and dull. To convince teenagers that Zap is an alternative to Coke and Fanta rather than simply a substitute for ordinary white milk, the Dairy Board created a promotion campaign based on the theme 'Zap . . . Get it and you've got it'.

Did that strategy work? Did the campaign really succeed in persuading teenagers that Zap is on a par with other soft drinks? To answer these questions, Philip Gendall (1986) designed a study that utilized discriminant analysis.

Application of discriminant analysis
About two years after the launch of Zap, 105 teenagers were asked to rate three of seven drinks – Coca-Cola, Fresh Up, Lemon & Paeroa, Fanta, Zap, lemonade, and milk – on 18 attributes. The ratings were on a seven-point scale with endpoints anchored as follows:

- Fizzy ... not fizzy
- Exciting ... unexciting
- Refreshing ... not refreshing
- Popular ... unpopular
- Modern ... old-fashioned
- Drunk mainly in a group ... drunk mainly on your own
- Drunk mainly in hot weather ... drunk mainly in cold weather
- Nourishing ... not nourishing
- Healthful ... unhealthful
- Filling .... not filling
- High in energy value ... low in energy value
- Natural taste ... artificial taste
- Strong flavour ... weak flavour
- Sweet ... not sweet
- A lot of added sugar ... no added sugar
- High in additives ... low in additives
- Luxury purchase ... everyday purchase
- Good value for money ... poor value for money

Exhibit 6.9 shows the mean ratings on 18 attributes for all products included in the study.

If the Dairy Board's positioning strategy had succeeded, the image of Zap would be similar to the image of Coca-Cola and the other carbonated soft drinks and well away from that of ordinary milk. An examination of the means (Exhibit 6.9) showed that the attributes that best described Coca-Cola, Fanta, lemonade, and Lemon & Paeroa were 'fizzy', 'refreshing', 'popular', 'drunk mainly in a group', 'drunk mainly in hot weather', 'high in additives', 'sweet', and 'containing a lot of added sugar'. White milk on the other hand, was regarded as unexciting and old-fashioned but nourishing, healthful, high in energy value, natural tasting, and free from additives. Fresh Up was seen as similar to milk, but not quite so natural and nourishing.

Zap appeared to fall somewhere between the carbonated soft drinks and the natural products (milk and Fresh Up). It was regarded as more modern than Coca-Cola, Fanta, lemonade, and Lemon & Paeroa, but less exciting, less popular, less refreshing, and less likely to be drunk in a group or in hot weather. Zap was considered to be more nourishing and healthful, higher in energy value, more natural tasting, and lower in artificial additives and sugar than the carbonated drinks, although it did not rate as well on these characteristics as milk or Fresh Up. Zap was more closely associated with Fresh Up and ordinary milk than with Coca-Cola and the other 'fizzy' drinks. In other words, the Board's attempt to position it as a soft drink had failed. But by how much had it failed, and just where was Zap positioned in relation to other drinks? To answer these questions Gendall applied multiple discriminant analysis to the data to compare the seven products *simultaneously* along the 18 attributes.

EXHIBIT 6.9

Mean ratings of seven drinks on 18 attributes

	Zap	Coca-Cola	Fanta	Lemonade	Lemon & Paeroa	Fresh Up	Milk
Fizzy	6.3	1.9	2.8	2.4	2.8	6.3	6.4
Exciting	4.6	3.7	4.2	4.1	3.3	4.6	5.6
Refreshing	4.1	3.2	3.1	2.8	2.6	2.6	3.8
Popular	4.4	1.9	3.4	3.0	3.2	3.3	3.2
Modern	2.6	3.4	4.0	4.4	3.1	3.5	5.9
Drunk mainly in a group	4.9	2.6	3.7	3.4	3.2	4.3	5.7
Drunk mainly in hot weather	3.6	2.4	2.4	2.4	2.2	2.7	4.4
Nourishing	3.3	5.4	5.4	5.1	5.1	2.6	1.8
Healthful	3.4	6.0	5.7	5.4	5.2	2.0	1.3
Filling	3.1	3.6	3.8	3.9	3.9	3.6	2.5
High energy value	3.4	4.0	4.6	4.4	4.5	2.7	2.2
Natural taste	4.3	5.6	5.3	4.5	4.4	1.9	1.6
Strong flavour	3.3	2.7	3.0	4.1	3.8	2.8	4.8
Sweet	3.2	2.6	2.0	2.6	2.7	3.6	5.9
A lot of added sugar	4.8	2.5	2.1	2.7	3.2	5.9	6.6
High in additives	3.4	2.2	2.0	2.4	2.9	5.2	6.2
Luxury purchase	3.4	3.7	3.7	3.8	3.6	3.3	6.7
Good value for money	5.3	4.5	4.8	4.4	4.7	3.8	1.7

Source: adapted from Fearon (1982).

Results of discriminant analysis

Multiple discriminant analysis can be used to *infer* the number and type of dimensions that underlie perceived similarities or dissimilarities. Multiple discriminant analysis also develops a geometric representation of the 'product space' such that products that are perceived as similar are positioned near one another and products that are seen as dissimilar are far apart. The objective of the exercise is to find the minimum number of dimensions of the set of objects compared and then to locate the objects in relation to these dimensions (and to each other).

The two-dimensional perceptual map created by discriminant analysis of attribute ratings and brands is shown in Exhibit 6.10. Attributes are shown as straight lines or vectors. The longer the line, the more important the attribute is in differentiating one product from another.

The perceptual map shows that Coca-Cola, Fanta, lemonade, and Lemon & Paeroa are closely grouped together in an area of the map that is described by the attributes 'sweet', 'exciting', 'refreshing', 'drunk in a group' and 'drunk in hot weather'. Milk and Fresh Up are found in the opposite direction, in an area described by attributes such as 'nourishing' and 'healthful'. Zap is positioned closer to Coca-Cola than to milk, but is obviously not regarded as a direct competitor of the 'fizzy' drinks. This suggests that the Dairy Board may have been successful in differentiating Zap from ordinary milk on some attributes, but that overall it was still regarded by teenagers as a flavoured milk rather than a soft drink.

EXHIBIT 6.10

Soft drinks perceptual map

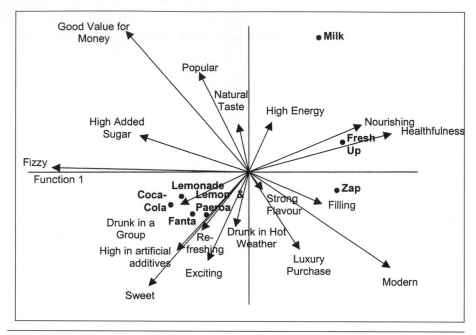

Product positioning map created using discriminant analysis (from Fearon, 1982).

Examination of the length and direction of the attributes displayed on the product positioning map also allows us to draw a number of conclusions about the relative importance of these attributes and the relationships between them. For example, because the line representing 'fizzy' is relatively long, we conclude that this attribute is more important in differentiating between the drinks studied than 'strong flavour', which has a very short attribute line. Since the lines representing 'nourishing' and 'healthfulness' are close together and point in the same direction, we can infer that teenagers see these attributes as very similar. On the other hand, 'good value for money' and 'refreshing' are at right angles, which indicates that there is no relationship between them; and 'good value for money' and 'luxury purchase' point in completely different directions, indicating that they are direct opposites as far as teenagers are concerned.

By drawing a perpendicular line from a particular drink to a particular attribute, it is possible to see how that drink compares with other drinks on that attribute. This is illustrated in Exhibit 6.11. From this diagram we can see that milk, for example, is regarded by teenagers as better value for money than Coca-Cola, which in turn is considered better value for money than Zap. However, while both milk and Coca-Cola are perceived as being above average value for money, Zap is seen as below average value for money (which is not surprising given that, at the time, a 500 ml carton of Zap was about 10 cents more expensive than a 333 ml can of Coca-Cola and more than twice as expensive as a 600 ml bottle of milk).

EXHIBIT 6.11

How an attribute relates to products

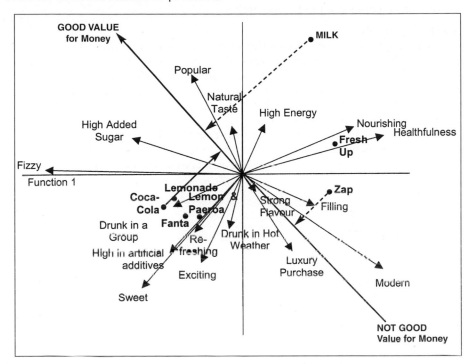

How discriminant analysis addressed the problem

The main conclusion of this study was that the Dairy Board's attempt to position Zap as a soft drink had failed. Does this mean it is impossible to reposition a milk product as a soft drink? Not necessarily. Many factors that are unrelated to the strategy itself might have contributed to the failure. Zap was the first cartoned liquid product sold in New Zealand, and it generated a great deal of controversy. This controversy was fuelled by arguments over the advantages and disadvantages of cartons, heightened by the fact that Zap was a milk-based product and hence subject to the political attention of the entire dairy industry. Critics of Zap attacked it on the grounds that it was overpriced, created litter, caused hyperactivity in children, and used imported packaging materials and hence would damage the New Zealand glass manufacturing industry. This made it very difficult for the Board to convince consumers that Zap was an alternative soft drink rather than simply another flavoured milk, and the successful positioning of Zap depended on achieving this objective.

Would the Dairy Board have achieved any success had it not had all these problems? Gendall concludes that this was probable since the Dairy Board had achieved some success with Zap. The Board had clearly differentiated Zap from milk and positioned it toward the exciting end of the soft-drink spectrum with no loss in perceived healthfulness compared to milk, as shown in the perceptual map.

On the basis of this observation and the other information derived from discriminant analysis, Gendall concluded that Zap should not try to compete directly with the carbonated soft drinks. A better strategy would be to seek a unique position in the market for Zap as an exciting and healthy drink.

Do purchasing criteria differ by country?

Marketing problem

Industrial purchasing agents from different countries may use different criteria when making purchasing decisions. Even within the same country, purchasing agents might apply different criteria to different products. Do UK industrial purchasing agents differ from US purchasing agents? If so, in what way? Is there any communality in the way they evaluate different products?

Application of discriminant analysis

To explore the differences between the two countries, Lehman and O'Shaughnessy (1974) asked 45 industrial purchasing agents (26 in the USA and 19 in the UK) to rate the importance of the following 17 criteria on a common scale:

- Reputation
- Financing
- Flexibility
- Past experience
- Technical service
- Confidence in salespersons
- Convenience in ordering
- Reliability data
- Price
- Technical specifications
- Ease of use
- Preference of user
- Training offered
- Training required
- Realibility of delivery
- Maintenance
- Sales service

Each purchasing agent was asked to do this separately for four different products. For each of the four products, Lehman and O'Shaughnessy applied a separate discriminant analysis to evaluate the difference between purchase agents of the two countries.

Results of discriminant analysis

Exhibit 6.12 shows the discriminant functions for four products (each derived separately from a two-group analysis). Group means indicated that the discriminant function had (arbitrarily) placed the UK on the positive side and the USA on the negative side. This meant that attributes with high positive coefficients were favoured by the UK agents while attributes with high negative coefficients were favoured by the US agents. Lehman and O'Shaughnessy were interested in large differences (identified by them to be those attributes with an absolute size of at least 1).

First, are the 17 criteria included in the study effective in distinguishing the two groups of agents? To evaluate this, Lehman and O'Shaughnessy examined the Mahalanobis D^2. To be significant at the 95% level the required Mahalanobis D^2 here is 27.8. All four Mahalanobis D^2 were substantially higher than this cutoff level, lead-

EXHIBIT 6.12

Discriminant functions: UK vs. US purchasing agents

Attribute	Product Type			
	I	II	III	IV
Reputation	−1.10	−0.16	−0.95	−1.02
Financing	−0.01	0.50	0.85	1.64
Flexibility	−0.19	0.53	1.07	−1.73
Past experience	−1.16	−0.11	−0.27	−1.05
Technical service.....................	0.19	2.38	1.57	−0.96
Confidence in salespersons	0.81	0.42	−0.55	0.48
Convenience in ordering..........	1.13	1.11	−0.01	0.18
Reliability data	−0.58	−0.67	−0.24	0.44
Price..	−0.81	−2.10	−0.21	0.27
Technical specifications...........	1.33	−0.66	−0.69	−0.30
Ease of use	−0.09	0.18	0.87	−1.17
Preference of user...................	−0.45	−2.02	−0.09	0.88
Training offered	0.15	−2.63	−0.96	−1.16
Training required	−1.55	0.39	−0.45	−0.12
Reliability of delivery	2.41	2.60	0.64	1.36
Maintenance	0.87	1.04	0.42	1.50
Sales service..........................	−0.39	1.03	1.30	1.96
Correctly classified (%)...........	84.4	86.7	77.8	84.4

ing to the conclusion that the variables included in the study are relevant in discriminating between the two sets of agents.

A quick examination of the results leads to two obvious conclusions. First, there are many large positive and negative coefficients, indicating that differences do exist between the UK and US purchasing agents. Secondly, the coefficients are not necessarily similar for the four products, indicating that the criteria applied are not necessarily identical in evaluating different products.

A closer examination of the discriminant functions reveals some interesting patterns. UK purchasing agents place greater emphasis on reliability of delivery and maintenance for all four products, on convenience for product types I and II and on sales service and financing for product types III and IV. The criteria used by US purchasing agents exhibit a very different pattern. They consider reputation as an important criterion for products I, III and IV, training offered for products II, III and IV, and price for products I, II. This leads to the generalization that UK purchasing agents are relatively more service-oriented while the US purchasing agents are somewhat more experience/reputation-oriented.

How well did discriminant analysis differentiate the UK agents from the US agents? Looking at the overall correct classification shown in Exhibit 6.12, the investigators concluded that they are large enough to support the conclusion that differences exist between purchasing agents of the two countries.

How discriminant analysis addressed the problem

Even though the study was based on a small sample, discriminant analysis showed that the purchase criteria used by UK agents are indeed different from those used by

US agents. It identified the patterns that underlie the differences in criteria as well. In addition, the analysis showed how the criteria used depend on the product type to be evaluated. The use of discriminant analysis also showed how agents strongly differ in their selection criteria for certain products (e.g., type II) and less strongly for certain other products (e.g., type III).

6.6 Caveats and concluding comments

Like regression analysis, discriminant analysis is easy to carry out and easy to analyse. As with regression analysis, a host of problems can affect the results. Most such problems are well documented and computer programs provide a number of diagnostics that would alert the researcher to their existence. Problems of regression such as collinearity, influential observations, non-linearity, etc. (discussed in the previous chapter on regression analysis), are relevant to discriminant analysis as well.

The number of cases correctly classified is the most commonly used criterion for assessing how well the model fits the data. In fact, most canned programs automatically calculate this model fit statistic. Many statisticians make the point that since the discriminant model was developed using the data set, evaluating the coefficients using the same data set would make the model appear more robust than it actually might be. To avoid this problem, as we noted earlier, one should consider using hold-out samples or jackknife procedures.

In many cases, the chance predictions of correct classification can be high depending on how many people belong to each group. For example, if 90% belong to group A and 10% to group B, by assigning everyone to group A, we can achieve 90% prediction accuracy. So it is important that we look more closely into impressive-looking results, and that we exceed the chance classification by a substantial margin before accepting the results as significant from a practical point of view.

Logistic regression may be used as an alternative to discriminant analysis if the normality assumptions that underlie discriminant analysis are not valid.

Bibliography

Further reading

For a detailed and exposition of discriminant analysis readers are referred to:

Huberty, Carl J. (1994) *Applied Discriminant Analysis.* Wiley Interscience, New York.

References

Dant, R.P., Lumpkin, James P. and Bush, Robert P. (1990) Private physicians or walk-in clinics: Do the patients differ? *Journal of Healthcare Marketing*, **10**(2), 25–35.
Fearon, D.C. (1982) A product positioning study using both metric and nonmetric multidimensional scaling techniques. Unpublished research report. Massey University, Palmerston North, New Zealand.
Gendall, P.J. (1986) The positioning of Zap: An application of discriminant analysis. In Richard J. Brook, Gregory C. Arnold, Thomas H. Hassard and Robert M. Pringle (eds), *The Fascination of Statistics*. Marcel Dekker, New York.
Huberty, C.J. (1984) Issues in the use and interpretation of discriminant analysis. *Psychological Bulletin*, 95, 156–71.

Lehmann, Donald R. and O'Shaughnessy, John (1974) Difference in attribute impor-
tance for different industrial products. *Journal of Marketing,* **38** (April), 36–42.
Manly, Bryan (1994) *Multivariate Statistical Methods: A Primer* (2nd edition).
Chapman & Hall, London.

7

Conjoint Analysis

7.1   What is conjoint analysis?

The problem of knowing what is really important to customers is a recurring one in marketing. How important are the different features in a video cassette recorder to a consumer? What features would a consumer give up for a 10% reduction in price? Would an investor give up broker recommendations in return for low commissions? For an airline passenger, is price more important than service? What is the combination of features that would be most appealing in a computer?

When we ask customers directly what they would ideally like, we make the situation somewhat unrealistic. We imply that there are no constraints on the choices. There is nothing that prevents a consumer from saying that every single feature presented is equally important. This offers little that is of value to the decision-maker. Again, two features rated as equally important by consumers may be incompatible. A consumer may rate both low price and luxury features in a car as equally important. Since luxury items cost more to manufacture, a manufacturer may not be able offer both low price and luxury features. In some cases, consumers may not be able to articulate precisely the importance they attach to different features of a product.

Yet consumers constantly trade off features that are less important to obtain features that are more important. Suppose you want to buy a computer and you have a choice between two models, one at £2400 and another at £3100. If this were the only consideration then the choice is obvious: the lower-priced computer is preferable. What if the lower-priced computer's keyboard is poorly laid out and the screen not as crisp as that of the higher-priced computer? What if the £2400 computer has a 3-month warranty while the £3600 computer comes with a 3-year warranty? Will you still choose the £2400 computer? If you are like most consumers, you are unlikely to decide on a purchase based on a single feature such as price. Rather, you will examine a range of features or attributes and then make judgements or tradeoffs to determine your final choice. Conjoint analysis, in its simplest form, makes the following assumptions:

- People ascribe measurable amounts of importance (called *utilities or partworths*) to different features of a product or service.
- When called upon to choose between alternatives, people simply add up the utilities for each feature of a given alternative.
- They will favour the alternative with the highest total utility.

The problem that the marketer faces is that consumers may be unable or unwilling to quantify precisely the importance they attach to different features of a product. Yet in real life, as the example above shows, they do make their choices based on the importance they attach to the features. Given this, can we infer the importance consumers attach to different features of a product from their overall choices? The answer is yes, given certain assumptions, and conjoint analysis is well suited for this purpose.

Conjoint analysis aims to estimate the importance a person attaches to different features of a product or service, without direct questioning. For instance, a builder may be interested in knowing the degree of importance a buyer attaches to features such as number of bedrooms, closeness to schools, type of neighbourhood and price. Respondents are not directly asked to specify the importance they attach to these features but are asked to indicate only their overall preferences to different houses with different combinations of features. From these overall judgements, conjoint analysis identifies the quantified importance (utilities) an individual attaches to different individual features.

By using conjoint analysis, we can determine the optimal features for a product or service, assess what products or services consumers will choose, and estimate the weight people will give to various factors that underlie their decisions. We can also use conjoint analysis to identify the best advertising message by isolating the features that are most important in product choice. Here are some typical problems handled by conjoint analysis:

1. A car manufacturer would like to know the ideal combination of features such as price level, fuel efficiency and safety equipment that will maximize sales.
2. A personal digital assistant (PDA) manufacturer would like to assess whether consumers would be willing to pay 50% for a new model that is lighter and capable of receiving emails.
3. An airline would like to know how many of the following a passenger will forgo to obtain tickets that are 10%, 20%, or 30% cheaper: legroom, in-flight meals, on-board newspapers, reservation and refund if ticket not used.

7.2   The conjoint analysis model

Conjoint analysis is best understood through a simple example. A PDA manufacturer would like to identify the best combination of the following three attributes: price ($250, $300, $350), weight (6 oz., 8 oz., 10 oz.) and colour (black and white, basic colours, all colours).

In conjoint terminology, the basic variables under consideration – in our example, price, weight and colour – are called *attributes* or *factors* (sometimes simply *variables*); the choices within an attribute – $250, $300, $350 for price in our example – are called *levels*. We are said to be dealing with conjoint analysis with three attributes, with three levels each. The importance weights we derive for any combination

EXHIBIT 7.1

All possible combinations of PDA attributes (3 attributes × 3 levels)

Model	Price	Weight	Colour
Model 1	$250	6 oz.	B&W
Model 2	$250	6 oz.	Basic colours
Model 3	$250	6 oz.	All colours
Model 4	$250	8 oz.	B&W
Model 5	$250	8 oz.	Basic colours
Model 6	$250	8 oz.	All colours
Model 7	$250	10 oz.	B&W
Model 8	$250	10 oz.	Basic colours
Model 9	$250	10 oz.	All colours
Model 10	$300	6 oz.	B&W
Model 11	$300	6 oz.	Basic colours
Model 12	$300	6 oz.	All colours
Model 13	$300	8 oz.	B&W
Model 14	$300	8 oz.	Basic colours
Model 15	$300	8 oz.	All colours
Model 16	$300	10 oz.	B&W
Model 17	$300	10 oz.	Basic colours
Model 18	$300	10 oz.	All colours
Model 19	$350	6 oz.	B&W
Model 20	$350	6 oz.	Basic colours
Model 21	$350	6 oz.	All colours
Model 22	$350	8 oz.	B&W
Model 23	$350	8 oz.	Basic colours
Model 24	$350	8 oz.	All colours
Model 25	$350	10 oz.	B&W
Model 26	$350	10 oz.	Basic colours
Model 27	$350	10 oz.	All colours

of features and levels are called *utilities* (the 'total worth' of that alternative). The weights we derive for each level of each attribute are called *partworths*. Partworths are utilities for a given attribute or a feature rather than for the alternative as a whole.

With three attributes and three levels, we can create 27 distinct PDA products (3 × 3 × 3 = 27), as shown in Exhibit 7.1. It is most likely that model 3 will be the most preferred and Model 25 will be the least preferred for most consumers. Other choices are not so clear-cut. Consumers are presented with various models and asked to rank them from the most preferred to the least preferred. (It is also not uncommon to ask the respondents to rate the alternatives instead of ranking them.) From these overall ranked preferences or ratings, conjoint analysis derives the importance consumers attach to the different levels of price, weight and screen.

However, this task of ranking 27 products is not an easy one for most consumers (even if they are willing to try). The number of alternatives can be reduced by using the *fractional factorial experimental design*.[1] This design is built into most computer programs which will identify the minimum number as well as the combinations. In our example, it turns out to be 9, as shown in Exhibit 7.2. Consumers are presented with these nine alternatives and are asked to rate each alternative or rank them from the most preferred to the least preferred.

EXHIBIT 7.2

Reduced set of aternatives chosen (using fractional factor design)

MODEL	PRICE	WEIGHT	COLOUR
Model 1	$250	6 oz.	All colours
Model 2	$350	8 oz.	All colours
Model 3	$300	10 oz.	All colours
Model 4	$300	8 oz.	B&W
Model 5	$250	10 oz.	B&W
Model 6	$350	6 oz.	B&W
Model 7	$350	10 oz.	Basic colours
Model 8	$300	8 oz.	Basic colours
Model 9	$250	8 oz.	Basic colours

Full profile vs. tradeoff

Full profile

The method of data collection described above is called the *full profile* or the menu approach, because we present each choice as a combination of levels of all features. The advantage of this approach is that it presents alternatives in a way that is closest to real life. Hence, it is a preferred method of presentation.

Pairwise tradeoff

The *pairwise tradeoff* method simplifies comparisons by presenting only two variables at a time. For instance, we might ask the consumer whether he prefers the $250 B&W model or the $300 all-colour one, and then whether he prefers the $300 10 oz. PDA or the $350 model that weighs only 6 oz. The rationale for this approach is that these bite-sized judgements are easier to make and the respondents are always focusing clearly on the feature being judged.

Which approach should you choose?

Although traditionally the full-profile approach has been the preferred method, it is generally unsuitable for telephone surveys. The *tradeoff* of two variables at a time is the most practical in that context. The tradeoff method has two major problems: first, it creates an artificial situation by comparing two features at a time, rather than looking at all features as a single package; and secondly, the number of comparisons can get too many, making it tedious for the respondent to answer the questions. For these reasons, the full profile is the more frequently used method.

---

[1]We assume that there is no interaction among attributes and levels. For instance, a consumer will attach the same importance to the addition of colour whether PDA costs $250 or £350. This means that we need to be concerned only with the main effects and not with interactions. It is because of this assumption that we are able to reduce the number of alternatives.

Utilities (partworths)

Traditional conjoint analysis assumes that the utility for an alternative j for consumer i can be derived from the following relationship:

$$u_{ij} = \sum_{k=1}^{K} \sum_{m=1}^{M} a_{ikm} x_{jkm},$$

where u_{ij} is the overall utility for an alternative j to consumer i,
$\quad\quad a_{ikm}$ is the partworth contribution or utility associated with the mth level
$\quad\quad\quad\quad$ of the kth attribute for consumer i,
$\quad\quad K$ is the number of attributes,
$\quad\quad M$ is the number of levels of attribute k, and
$\quad\quad x_{jkm}$ is the dummy variable value (1 if present, 0 if absent) of the mth level
$\quad\quad\quad\quad$ of the kth attribute.

Several different procedures are available for estimating the traditional model. The aim in all cases is to estimate a_{ikm} such that the derived u_{ij} will closely match the consumer's preference ranking.

- If the consumer's evaluations are ratings, then ordinary least-squares (OLS) regression can be used with these ratings as the dependent variable. The attributes (independent variables) are treated as dummy variables for attribute levels. It is also common to code dummy variables using effects-coding procedures. The resulting regression coefficients are usually rescaled (see below) and these rescaled coefficients are the conjoint utilities.
- If the consumer's evaluations are rankings, then OLS may be problematic since the technique assumes the dependent variable to be at least interval-scaled. In such cases other procedures such as monotonic analysis of variance (Kruskal, 1965) and linear programming (Srinivasan and Shocker, 1973) are more appropriate. Monotonic ANOVA is achieved through two iterations. First, the importance of attributes is estimated by OLS. Then a monotone transformation is applied to the predicted preferences to make their rank order conform to that of observed ranks. The monotone transformation amounts to averaging the pairs of predicted values for which the orders are different compared to the corresponding pair of observed values. These two steps are alternated until convergence (to a specified criterion) is achieved. Other alternative techniques for ranked data include rank-order logit (Chapman, 1997).

However, many researchers have been finding that OLS applied to rank-ordered data gives utilities estimates that are not too different from those obtained using more exact procedures (Cattin and Wittink, 1976; Carmone, Green and Jain, 1978). Consequently, effects-coded dummy variable regression is being widely used for both ranked and rated responses.

If we apply dummy variable regression to the conjoint problem, we have

$$R_{ij} = \sum_{k=1}^{K} \sum_{m=1}^{M} a_{ikm} x_{jkm} + \epsilon_{ij},$$

where ϵ_{ij} is the error term, assumed to be normally distributed with zero mean and unit standard deviation for all i and j.

Importance of an attribute

The importance I of a given attribute k is estimated in terms of the range of its part-worths. A larger range indicates that the attribute exerts a greater influence on the dependent variable. The importance of an attribute can be specified as:

$$I_k = \max{(a_{km})} - \min{(a_{km})}, k = 1, ..., K.$$

To assess the importance of a given attribute in relation to others, we can normalize it:

$$w_k = \frac{I_k}{\sum_{k=1}^{K} I_k}$$

so that $\sum_{k=1}^{K} w_k = 1$.

Segmenting the market

Conjoint utilities are estimated for each individual separately. While we can aggregate the utilities for the sample as a whole, at times such aggregation may mislead. In our PDA example, it is quite possible that there are consumers for whom the utility for price is high and consumers who would rather have a lighter PDA, irrespective of the marginal cost. When we aggregate, high utility for price for one group of consumers may be cancelled out by the low utility for price for another group. It might look as though price is not a major consideration, while in reality the market may be segmented: the manufacturer may be able to market two models – a basic one at a low price and a lighter one at a higher price – rather than a single model that may have low appeal to both groups.

Consequently, whenever there is reason to believe that consumers' needs may not be homogeneous, it common to conduct cluster analysis (see Chapter 4) on the utilities to identify consumers whose utilities are similar.

Estimating the market share: Probability and first-choice models

The results of the conjoint analysis may be extended to predict a product's market share. A number of models with somewhat differing underlying assumptions are available for *market simulation* analysis.

Probability models

If we hypothesize that the probability of a consumer buying a product can be inferred from the proportion of the total utility the consumer assigns to that alternative, then the probability of purchase (p_{ij}) can expressed as

$$p_{ij} = \frac{u_{ij}}{\sum_{k} u_{ij}}, \text{ for } j \in \{1, ..., J\}$$

where u_{ij} is the utility estimated for customer i for product j. Market share for a given alternative j (MS_j) can then be calculated by

$$MS_j = \frac{\sum_{i=1}^{I} v_i p_{ij}}{\sum_{j=1}^{J} \sum_{i=1}^{I} v_i p_{ij}},$$

where I is the number of customers in the study, J is the number of alternative products in the study, and v_i the relative volume of purchases made by customer i, with the mean volume across all customers indexed to 1. There are also other models (such as logit choice) which use this basic share of utility.

First-choice models

An alternative to the probability model is the first-choice model, also known as the maximum utility model. This 'winner-take-all' approach holds that a customer will choose the one alternative that has the highest utility and reject all others. This resembles a realistic purchase occasion model since, on any given occasion, a customer is likely to choose one among the many alternatives. However, it is less realistic in contexts (such as fast-moving consumer goods markets) in which a consumer is likely to buy more than one brand, especially over a period such as a year (Ehrenberg, 1988).

Which model to use?

Probability models suffer from a major weakness. Since the probabilities are dependent only on the alternatives included in the study, the probability of purchase can be arbitrarily changed by adding a couple of alternatives that a consumer is unlikely buy but will have some utility value nevertheless. Another disadvantage of these models is that they are sensitive to scale ranges within which utilities are measured; they are also affected by certain types of linear transformation.

First-choice models, on the other hand, tend to predict extreme market shares. They also tend to be unstable especially when there are alternatives that are about as attractive as the first choice. In such cases, any minor bias or even some random component can make one product look far more attractive than it really is, since these models do not distinguish between strong and weak preferences of a consumer.

Since all models have their strengths and weaknesses, some formal rules have been proposed to decide on the best model for any given context. The *alpha rule* proposed by Green and Krieger (1993) suggests that we first compute the market shares of the existing products only, compare them with the actual share of these products using different models and accept the model that comes closest to the known shares. The *randomized first-choice rule* suggested by Orme and Huber (2000) extends the alpha rule by assuming the existence of a random component in translating utilities into choice.

How good are the conjoint results?

Face validity

To validate the results generated by conjoint analysis, we first examine the utilities at the aggregate level. We first should look for *face validity*. For instance, in general, the utility should decrease inversely with price and directly with quality. Are the results in the expected direction? If not, are the results readily explainable in the present context?

Correlation between actual and obtained ranks

Once we establish face validity, we need to check the fit of the model to the actual data. Since conjoint analysis derives utilities on the basis of customer preferences, can we go back and reproduce customer preferences on the basis of derived utilities? To do this, we correlate the actual ranking (or rating) of different alternatives by the consumers with the rank of utilities for the same alternatives, as derived through conjoint analysis. We do this for each consumer (since utilities are derived for each consumer separately) through the use of Kendall's tau, Spearman's rho (for ranked data) or Pearson's *r* (for ratings data).

Holdout stimuli

Since validating a model using the same data as were used to create the model is not generally considered to be a good procedure, the results are often validated by means of 'holdout stimuli'. Here the researcher has more alternatives than are necessarily evaluated by consumers. Once the utility estimates are obtained using only the ones called for by the experimental design, these parameter estimates are applied to the holdout alternatives. The ability of these parameter estimates to correctly reproduce the preference ranking (or rating) of consumers is a measure of their validity.

7.3   Conjoint analysis: computer output

A financial institution is about to issue a new bond. To ensure competitive advantage, the institution wants to add 'bells and whistles' such as more frequent interest payments and higher interest rates to its offer. The alternatives considered are shown in Exhibit 7.3. A conjoint analysis survey was administered to find out how attractive these features are to potential buyers. In all, 500 potential investors were interviewed.

EXHIBIT 7.3

Attributes and levels for a new bond issue

Attributes	Levels			
Interest rate	6.0%	6.5%	7.0%	7.5%
Term	1 year	3 years	5 years	
Withdrawal penalty	1.5%	2.0%	2.5%	
Payment frequency	Semi-annual	Quarterly	Monthly	

If we used all possible combinations, we would have 108 possible combinations (4 interest rates $\times$ 3 terms $\times$ 3 withdrawal penalties $\times$ 3 frequency of interest payments = 108). Because 108 choices would be too many for people to rank, they were reduced to 16 choices using the fractional factorial design. These 16 choices were then printed separately on 16 separate cards (Exhibit 7.4 shows two examples of such cards). Customers were asked to rank these 16 choices.

The results of the study were submitted to a conjoint analysis program using the Monotonic ANOVA approach, which is presented in Exhibit 7.5. (A number of commercial conjoint programs are available on the market. Different programs provide different types of output and graphic capabilities. While they may all look different, most good commercial programs include analytic output similar to the one presented here.)

Individual utilities or partworths

Conjoint analysis works at the individual level. This means that utilities for each individual are derived separately, as shown in the computer output. Individual utilities are no more useful than individual ratings of a product. To get a clearer picture of the

EXHIBIT 7.4

Examples of cards used in the bond study

OPTION J
Face Interest Rate	Fixed Rate: 7.5%
Interest Cheques	Quarterly
Term to Maturity	3 Years
Early Redemption	Can be cashed at any time. If cashed before maturity, Early Redemption Rate would be 0.5% lower than the fixed interest rate.

OPTION M
Face Interest Rate	Fixed Rate: 8%
Interest Cheques	Semi-annually
Term to Maturity	5 Years
Early Redemption	Can be cashed at any time. If cashed before maturity, Early Redemption Rate would be 2.5% lower than the fixed interest rate.

EXHIBIT 7.5

Conjoint analysis output, annotated

```
001    0.39 0.41 0.38 0.69 0.68 0.43 0.38 0.69 0.68 0.68 0.71 0.24 0.11
002    0.39 0.41 0.38 0.69 0.68 0.43 0.38 0.69 0.68 0.68 0.71 0.24 0.11
003    0.39 0.41 0.38 0.69 0.68 0.43 0.38 0.69 0.68 0.68 0.71 0.24 0.11
004    0.39 0.41 0.38 0.69 0.68 0.43 0.38 0.69 0.68 0.68 0.71 0.24 0.11
```

These are 13 partworths corresponding to 13 attribute levels shown in Exhibit 7.3. (Shown only for the first four respondents. Some programs may not output this part.)

Respondent ID

```
Kendall's Tau (Partial output)
    001    0.82
    002    0.94
    003    1.00
    004    0.79
```

The first column shows the respondent ID. The second column shows the correlation between the respondents' original ranking of the alternatives and the ranking of utilities for each alternative. For example, for respondent 4, the derived utilities for each of the 16 alternatives presented were in the same rank order as the ones given by the respondent for those alternatives.

EXHIBIT 7.6

Average utilities for different levels

Means				
Interest rate	6.0%	6.5%	7.0%	7.5%
Utilities	−0.35	−0.10	0.10	0.35
Term	1-year	3-years	5-years	
Utilities	0.10	0.00	−0.10	
Withdrawal penalty	1.5%	2.0%	2.5%	
Utilities	0.30	−0.10	−0.20	
Payment frequency	Semi-annual	Quarterly	Monthly	
Utilities	−0.10	0.0	0.10	

importance of different attributes and levels, we examine the average utilities across all respondents in the output (Exhibit 7.6). The results appear to make intuitive sense. For instance, we would expect people to have higher utility for higher interest rates; we would expect them to have lower utility for higher penalties. Results are in the expected direction, which tells us that probably the model is appropriate for our problem. Exhibit 7.6 can be used to assess the face validity of the results.

Finding the relative importance of attributes

To understand the relative importance of the different attributes, we may plot the mean utilities for different levels of attributes, taking care that the *y*-axis has the same value for all attributes. When we examine such plots (Exhibit 7.7), we note that interest rate has a very steep line, meaning that interest rate strongly influences utility – the higher the interest rate, the greater the utility. On the other hand, the line for frequency of payment is fairly flat, meaning that the frequency of interest payment changes the utility only slightly and that it is of little importance to the consumer. In other words, the range of utilities for a given variable shows its relative importance.

To quantify the impact of an attribute, we first calculate the range of utilities for each attribute by finding the distance between the lowest partworth and the highest

EXHIBIT 7.7

Attribute elasticities

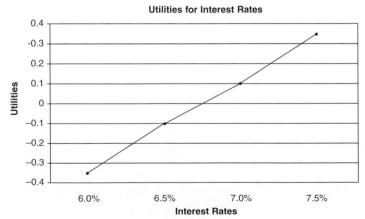

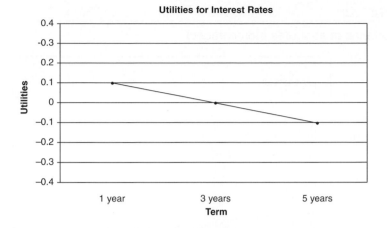

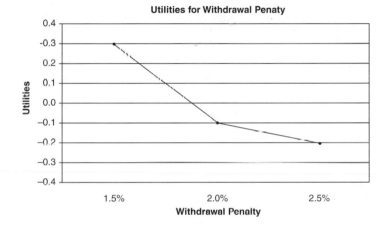

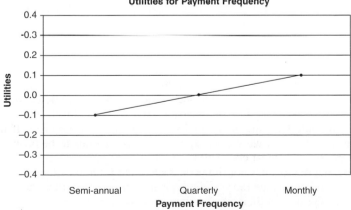

EXHIBIT 7.8

Importance of attributes (percentages)

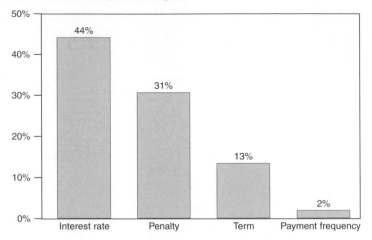

partworth and adding them across all attributes. This works out to 1.60 in our example (see Exhibit 7.6). The relative importance of an attribute is the range of utilities for that attribute as a percentage of the total of all ranges. As an example, the utility range for interest rate is 0.70. This represents 44% of the sum of ranges (0.70/1.60 = 44%). Hence the relative importance of interest rate is 44%. Exhibit 7.8 shows the importance of the attributes included in this study.

Assessing the relative value of the proposed products

Since we have utilities for each attribute level, we can obtain the total utilities for any proposed product. In our case, if we have no restriction on what is feasible, we will have potentially 108 products. By combining the utilities for the component features, we can arrive at the total utility for any combination of features. However, not all combinations of attributes may be feasible from a manufacturing or marketing point of view. If only about 10 of these will make sense from a marketer's point of view, then the marketer could choose to market the product that will maximize the market share or maximize profit. But before any such decisions are made, we have to make sure that the results are valid. For this we use correlation.

Correlations

While the average utilities can be used as an intuitive measure of validity, we also need a more objective measure. One common measure of validity is the correlation between the ranks given by a respondent and the ranks predicted by the utilities derived by conjoint analysis. This measure tells us how closely the derived utility measures correlate with the actual ranks given by each respondent. If the conjoint model fits the data, then we would expect most correlations to be positive and high, say, +0.8 or above. If most correlations are low or negative, then we can reasonably assume that the model does not fit the data adequately. On the other hand, if there are only a few cases for which the correlation is low or negative, then these cases can be treated as exceptions (outliers) and perhaps may be eliminated.

Market simulation

Many computer programs are capable of computing market shares of different alternatives. The last part of the output shows the market share of different alternatives using different methods. These results involve assumptions about consumer behaviour and therefore should be treated with caution, and perhaps should be validated using other models, as discussed earlier.

7.4   Conjoint analysis: marketing applications

How to understand consumer preferences

Marketing problem

Consumer decisions with regard to the choice, purchase and use of products are of critical interest to marketers. Such decisions can become highly complex as the number of alternatives increases. Further complexity is added to this problem since a number of attributes characterize each alternative.

Consequently, when the sales of a product fall, it is not always easy to pinpoint the reasons for the decline. For instance, consider the problem of falling sales of its chewing gum brand faced by a multinational company in the Sri Lankan market (Gunaratne, 2001). It is important for the company to take steps to slow down or reverse such a fall. To do this, the company has to understand the causes that led to the decline in sales as well as the reasons why consumers prefer one gum over another.

Application of conjoint analysis

To understand the reasons for the decline in sales, the company interviewed 300 consumers, 76% of whom believed that long lasting flavour influenced their purchase decisions, suggesting that the sale of any brand of chewing gum is influenced by the length of time the flavour is perceived to last in the mouth. The respondents were asked to judge samples of chewing gum with different flavour levels (high, medium and low, where 'high' means longer lasting). While the results showed that 82% preferred the high flavour product, market analysis revealed that none of the leading brands were judged to have higher flavour levels than competing brands.

The company decided to carry out a conjoint analysis to address the problem. Attributes and levels to be included in the conjoint study (see Exhibit 7.9) were decided on the basis of a number of focus groups. The number of attributes and levels shown on Exhibit 7.9 produce a combination of 72 alternatives (4 × 3 × 3 × 2). A fractional factorial design was used to reduce the number of alternatives. Exhibit 7.10 shows two of the profiles used in the study. The profiles were presented to respondents at random who then ranked the alternatives from the most preferred to least preferred. Partworths for attributes and levels were derived from these ranks.

Results of conjoint analysis

The relative importance of attributes and partworths of levels are shown in Exhibit 7.11.

Price. While the results of the initial survey had shown that 82% of the respondents preferred chewing gum with a longer-lasting flavour, the overall results of the conjoint analysis showed that price exerts an even greater influence on preference (relative importance 41%). An interesting finding was that the influence of price was not

EXHIBIT 7.9

Attributes and levels

Attributes	Levels			
	1	2	3	4
Country of origin	USA	UK	Malaysia	India
Price (Sri Lankan rupees)	3.65	3.15	2.75	
Flavour	High	Medium	Low	
Packaging	Silver	Wax		

EXHIBIT 7.10

Two examples of profiles used in the study

PROFILE 1
Country of origin	UK
Price per pack	SLR 3.65
Level of flavour	High
Packaging	Silver foil

PROFILE 2
Country of origin	Malaysia
Price per pack	SLR 3.15
Level of flavour	Medium
Packaging	Glossy waxed paper

EXHIBIT 7.11

Attribute importance and partworths

Attribute/level	Relative importance	Partworth
Price per pack	41%	
SLR 3.65		−0.02
SLR 3.15		+0.17
SLR 2.75		−0.15
Level of flavour*	39%	
High		+0.53
Medium		+0.93
Low		−0.61
Country of origin	11%	
USA		+0.09
UK		+0.24
Malaysia		+0.11
India		−0.44
Packaging	10%	
Silver foil		−0.38
Glossy waxed paper		+0.38

*These partworths are as reported in the original paper. I believe that there is a calculation error for this attribute, since the partworths do not add up to 0, as they do for other attributes.

linear for chewing gum. Consumers tended to avoid both low and higher prices and preferred to pay SLR 3.15, the mid-range price.

Flavour. The second most important attribute that influences consumer preference was flavour, with a relative importance of 39%. The relationship between flavour and preference also turned out to be non-linear, with consumers preferring medium flavour levels to high or low flavour levels, again contrary to what they had stated in initial research.

Country of origin. There is a distinct preference for chewing gums from the UK, with those from India being the least preferred, though this attribute has a relative importance of only 11%.

Packaging. Like country of origin, packaging exerted a low relative influence (10%) on consumer preferences. The results indicated a preference for the glossy waxed paper. Gunaratne hypothesized that the preference was perhaps the result of an association between the use of silver foil and expensiveness.

How conjoint analysis addressed the problem

The stated preference of consumers was for chewing gum with longer-lasting flavour. Without the conjoint analysis study, this preference would have been the basis for product improvement. Since flavour accounts for the highest cost component of the product, even a marginal increase in the level of flavour leads to substantial cost increases. While the product would have delivered what consumers had said they wanted, it would have added to the price of the product.

Conjoint analysis showed that such a strategy would have failed on two counts. First, it indicated that price had a greater utility than flavour, which meant that, while consumers preferred lasting flavour, they were not necessarily willing to pay more than a certain amount for it. Secondly, 'long-lasting' did not mean for ever or even for a long time. The consumer wanted flavour to linger, but perhaps not for too long.

Although, presumably, the company may not have much control over the country of origin, this attribute exerted much less influence on preference, as did the packaging.

While conjoint analysis provides input for marketing decisions, Gunaratne warns that the market performance of products is also influenced by the marketing-mix variables such as advertising, promotion and distribution, which affect the awareness and the distribution of choice alternatives. The effects of these variables, which are not captured in the study, limit its predictive validity of market success.

What is the best package to offer for a sports event?

The marketing problem

Correctly pricing a product or a service is one of the most important problems faced by a marketer. Pricing a product or service too high might result in lost sales; pricing it too low might result in lost revenue if the consumer would have paid more. Conjoint analysis is often used to determine the appropriate price for combinations of products and services, as is shown in this case of the Australian Indycar Grand Prix.

The 1993 Australian FAI Indycar Grand Prix was a four-day event, featuring Indycars, Nascars, Auscars, Sports and Transam, and drag racing, as well as an extensive programme of ancilliary events, including aerial displays, a Mardi Gras, jazz concerts, beach parties, a rock concert and a Miss Indy competition. This strong mix of attractions posed a challenge: how to optimize the revenue while keeping the attendance as high as possible?

Application of conjoint analysis

To address the marketing problem, MacLean and Croft (1993) carried out a series of face-to-face interviews with 202 Brisbane residents (18–55 years of age) who had not previously attended the Gold Coast FAI Indy, but had an interest in doing so. Conjoint analysis was applied to understand the following issues:

1. *Advance purchase discount.* Should the promoters issue advance tickets that carried discounts? If so, what price–discount combination?
2. *Transportation.* Should the promoter arrange for free transportation? If so, what type?
3. *Souvenir pack.* Should the promoter offer souvenir packs to those attending? If so, for what value?

The chosen alternatives are shown in Exhibit 7.12. There were a total of 36 (4 × 3 × 3) possible combinations, and these were then reduced to 16 using a fractional factorial design. Respondents scored each option in order of preference.

EXHIBIT 7.12

Attributes and levels used in the study

Attribute	Levels
Advance purchase discount	No discount, 10% (1 month in advance), 20% (2 months in advance) and 30% (3 months in advance)
Transportation	No free transport, free bus transport, free limousine transport
Souvenir pack	None, $25 worth, $50 worth

Results of conjoint analysis

From the preference scores provided by the respondents, partworths were derived for each level of each attribute, as shown in Exhibit 7.13. A quick examination of the exhibit reveals some interesting patterns. The overall effect of discounts is much more pronounced than either free transportation or free souvenirs: for advance purchase discounts, the partworths range approximately from −20 to +15, a range of nearly 35 overall, compared to a range of 20 for free transportation and 12 for free souvenirs. In fact, discount is more important than the other two attributes put together. This elasticity is expressed in percentage terms in Exhibit 7.14. It is interesting to note that, while discount is important to all age groups, younger people are willing to trade off some discount to obtain free transportation, presumably because they may not have cars. The oldest group (40–54) attaches most importance to discount.

Although conjoint analysis showed the relative importance of attributes and levels, it did not show the absolute interest in attending the 1993 Grand Prix. This information was obtained thorough straightforward questioning of interest. If advance purchase discounts were included, 72% considered themselves interested in attending, compared to 58% at the preliminary price. The proportion that would still be unlikely

EXHIBIT 7.13

Partworths of different alternatives*

Advance purchase discount	
0% discount (No advance purchase)	−20.0
10% discount (1 month in advance)	−3.0
20% discount (2 months in advance)	+8.0
30% discount (3 months in advance)	+15.0
Transportation	
No free transport	−12.6
Bus transport free	+5.2
Limousine transport free	+7.4
Souvenir pack	
No Free Souvenir Pack	−7.0
$25 Free Souvenir pack	+1.5
$50 Free Souvenir pack	+5.5

*These values are approximate, derived from measuring the lengths of bars in a bar chart presented in the original paper. The original paper did not provide the numerical values.

EXHIBIT 7.14

Relative importance of attributes
(as measured by their relative elasticity)

		Age groups		
	Total (%)	18–24 (%)	25–39 (%)	40–54 (%)
Advance purchase/ticket discount	52	42	52	67
Free return transport	30	37	30	18
Free FAI Indy souvenir pack	18	21	18	15

or very unlikely to attend fell from 26% at the preliminary price to 16% with some form of price discounting.

How conjoint analysis addressed the problem

The conjoint analysis (along with other survey questions) showed there was a significant interest in attending the FAI Indy Grand Prix, but that the preliminary prices for each ticket type were generally considered too expensive without added inducements to attend. Conjoint analysis specifically showed that discount is a powerful factor in attracting a potential audience.

The fall in interest in the FAI Indy event (which presumably necessitated this study) was reversed with the introduction of advance purchase discounts. The introduction of levels of discounts up to three months in advance was used to improve the attractiveness of the FAI Indy ticket prices. The conjoint analysis also revealed the higher relative importance of free transportation to the younger audience. The difference in interest between limousine and bus travel was minimal, further confirming the hypothesis that transportation was perhaps considered essential by the younger group.

Pricing research had indicated that the optimum price for an FAI Indy Grand Prix ticket was between $45 and $55. The inclusion of a 30% discount to the preliminary ticket price 3 months in advance would make the optimum price achievable. For weekend flexiseats (one of the packages for attendance), the inclusion of three-tiered pre-purchase discounts enabled the organizers to increase the preliminary price from $100 to $110 a ticket, since those interested in buying at a favourable price would do so through advance purchase. Similarly, for weekend grandstand reserved (another package) the discounts enabled the preliminary ticket price to increase from $120 to $150 per ticket.

The application of conjoint analysis in this instance enabled the organizers to increase the price of admissions while at the same time maximizing the attendance by offering discounts.

How to assess transportation preferences

Marketing problem

A problem that many local governments face is that of assessing consumer transportation preferences. There are many elements that decide transportation choices: travel time, housing density, air quality, taxes, uses of open space and the impact of new road construction. These elements are good candidates for tradeoff analysis. For instance, new road construction may be perceived to be desirable, but will consumers pay more in taxes to make it happen?

In 1991, the Canadian city of Calgary faced the problem of assessing its citizens' attitudes and preferences to several elements of transportation (Hunt, Abraham and Patterson, 1995). In particular, the city needed to understand a range of specific tradeoffs of 19 different elements affecting urban transportation so it could design its transportation system in line with citizens' preferences.

Application of conjoint analysis

There were many elements that were hypothesized to affect transportation preferences. These included mobility, air quality, residential density, use of environmental space and funding sources such as taxes and user charges. The attitudes and preferences of Calgarians with regard to four areas of transportation and housing types were of interest for this project:

1. mobility (travel times to work for both auto and public transport modes)
2. built form (housing types, ranging from single family dwelling to highrise)
3. environment (frequency of noticeably bad air quality, proximity to an environmentally significant area – ESA – such as a river or valley, loss of open space because of road construction through an ESA)
4. costs and taxes (auto, public transport)

This yielded several elements or levels, for all attributes combined. Each of these elements had numerous possible states. The state for each element in an alternative was selected randomly according to a uniform distribution among the possible states. Illogical alternatives were excluded from the alternatives presented to the respondents: for example, if the home location was not near an ESA, it could not be allowed to be near new road construction within an ESA.

The hypothetical scenarios were created by combining randomly selected levels for each attribute as described above. Respondents were asked to rank these hypothetical scenarios in order of preference as future home locations based on the levels and attributes for each alternative.

Hunt *et al.* carried out 961 in-home, face-to-face interviews, using random sampling procedures. As a part of the study, they also collected demographic and socio-economic information. Each respondent completed three separate rounds of ranking. Hunt *et al.* felt that this improved the efficiency of the data collection – second and third rounds were considered likely to provide more accurate indications in that they were performed by 'experienced' respondents. The maximum number of rankings was set at three in an attempt to keep the interview under 30 minutes. Four different hypothetical situations were ranked in each round, thereby providing six pairwise comparisons per round. More than four situations would have provided more comparisons, but the ranking task would have been more difficult for respondents.

The descriptions for each alternative were presented on separate sheets of paper, allowing respondents to physically arrange the alternatives in order of preference. Of the 11 elements available, only nine were presented in a given interview. An interview considered either a river valley or an ESA, not both. The element for proximity to open space not considered and the element for road construction through or over the open space not considered were also not presented in an interview.

To reduce complexity, the number of elements varying in a given ranking was kept to five by holding the four randomly chosen elements constant across the four alternatives. Each of these four was held constant at a state selected randomly, while the other five were allowed to vary among the alternatives. The order of presentation of the five varying elements was selected randomly for each interview by the computer program. This order was maintained for all rankings in a given interview.

Results of conjoint analysis

Exhibit 7.15 shows how changes in conditions regarding each of the elements impact the typical household, based on the utilities for different level of attributes. The trade-off between auto drive time and costs for these results implies the value of auto drive time for a work trip is $11.06 per hour. A similar comparison for the value of transit time to work implies a value of $6.95 per hour. These findings were consistent with an earlier study (Hunt, 1992).

Utilities for different housing types show that all other types of housing are less desirable than the single-family dwelling. Those for the infill, duplex, and townhouse categories are approximately 4 times as large as that for a 10-minute auto drive time to work, i.e., an average Calgarian would rather drive an additional 40 minutes to work than live in one of the denser housing types. Clearly, housing type and density are very important to Calgarians.

An increase in taxes by $100 per month is perceived as the rough equivalent of switching from a single family dwelling to a townhouse. A one-minute further trip to work is regarded approximately the same as $2.50 in municipal taxes per month.

Bad air quality going from never bad to bad once per week is approximately equal to $100 per month of municipal taxes. Being close to a river valley or an ESA is seen as a good thing. Disturbing these areas with major road development is seen as positive if the road crosses a river valley and negative if it passes through an ESA. Road development through ESAs far from the home location is similar to adding a 8.6 minutes to driving time to work.

How conjoint analysis addressed the problem

The study showed through utility estimation what is important for Calgarians and put a financial value on it. For instance, it showed that

EXHIBIT 7.15

Utilities and standard errors of different elements

	Partworth	Standard error
10 minutes' auto time	−0.2840	±0.0133
$1 auto cost	−0.1541	±0.0180
10 minutes' transit time	−0.1684	±0.0111
$1 transit cost	−0.1453	±0.0335
Single family	0.0000	0.00000
Infill	−0.8969	±0.07260
Duplex	−1.0460	±0.07160
Townhouse	−1.1560	±0.07310
Medium density	−2.1270	±0.07799
Highrise	−2.5520	0.08180
$100 tax per month	−1.1260	±0.03530
Air never bad	0.0000	0.0000
Air bad 1 day per year	−0.2634	±0.0543
Air bad 1 day per month	−0.6724	±0.0563
Air bad 1 day per week	−1.2330	±0.0590
River far	0.0000	0.0000
River near	+0.4967	±0.0599
ESA far	0.0000	0.0000
ESA near	+0.6820	±0.0654
No major road over river	0.00000	0.0000
Major road far over river	+0.20640	±0.0581
Major road near over river	+0.05002	±0.0785
No major road through ESA	0.00000	0.0000
Major road through ESA	−0.24520	±0.0647
Major road near through ESA	−0.36630	±0.0826

This table was kindly provided by J. Douglas Hunt and printed with his permission.

- a one-minute further trip to work is regarded as approximately the same as $2.50 in municipal taxes per month;
- bad air quality going from never bad to bad once per week is approximately equal to $100 per month of municipal taxes.

These tradeoff rates provide valuable input to the decision-maker when considering whether municipal tax dollars should be spent on transportation improvements to provide travel time savings. The tradeoff utilities also show that Calgarians are willing to pay as much as $100 in taxes to have clean air. By attaching monetary value to alternative decisions, conjoint analysis provided actionable information to the city to enable it to make decisions that are consistent with consumer transportation preferences.

7.5   Caveats and concluding comments

Conjoint analysis, as noted earlier, is one of the most frequently used multivariate techniques in marketing. So it is important to know its limitations. The limitations of conjoint analysis are less visible than those of many other techniques. For instance, in regression analysis, problems such as multicollinearity, non-linearity, and influential observations are well defined and can be identified (and possibly corrected) using statistics available from the computer outputs. The problems with conjoint analysis tend to be more conceptual. A less sophisticated user may not even be aware of the existence of some of the problems.

Attributes and levels

Although for illustrative purposes we used the aggregate data, the utilities are calculated at the individual level, except when we use discrete-choice models. We should also keep in mind that this is only a convenient way of expressing the importance of a variable and we should not over-interpret it. Studies indicate that the utility range is related to the number of levels within variable. Furthermore, it may be tempting to make generalizations such as 'customers do not care about the penalty for early redemption'. Such inferences ignore the fact that the results can be valid only for the levels tested and may not be generalized. For instance, while customers may not be bothered by a 2% penalty in a conjoint study, if we had included a penalty level of 3% in the same study it is quite conceivable that it might have altered the relative importance of attributes. Penalty could have emerged as an important variable, perhaps even more important than the attribute currently considered most important.

Interaction effects

The general conjoint model assumes that partworths are additive. This means that a given level of an attribute has the same partworth, irrespective of the context. For instance, if interest rate and penalty for early withdrawal for term deposits are included in a conjoint study, it assumes that penalty for withdrawal will have the same partworth whether the interest rate is 2% or 10%. This assumption may not always be tenable. It is reasonable to expect that higher interest rate will increase the tolerance for penalties. The general conjoint model does not include interactions.[2]

Number of levels within an attribute

There is some evidence to show an attribute with a larger number of levels tends to be more elastic than one with a fewer number of levels. So it is preferable to keep the number of levels the same for each attribute. However, this may not always be possible in practice.

The best alternative may not appeal to the consumer

When we ask a consumer to rank a number of alternatives, we assume that the respondent *likes* his or her first choice. In reality, the first ranked alternative may be completely unacceptable to the consumer. It was ranked first only because it was the 'best' of the alternatives presented to the respondent. There are other models – such as choice-based conjoint analysis – which address this issue.

---

[2]There are models that include interaction effects, but they are not frequently used in commercial research.

The longevity and widespread appeal of conjoint analysis to marketing researchers indicate that in practice it provides useful input to marketing researchers, despite its various, well-documented limitations.

Bibliography

Carmone, Frank J., Green, Paul E. and Jain, Arun K. (1978). Robustness of conjoint analysis: some Monte Carlo results. *Journal of Marketing Research*, **15**, 300–3.

Cattin, Philippe and Wittink, Dick R. (1976) A Monte Carlo study of metric and non-metric estimation methods for multiattribute models. Working Paper no. 341, Stanford University.

Chapman, R. G. (1997). Exploiting the information content of ranked choice sets with homoscedastic and heteroscedastic multinomial logit models. Paper presented at the American Marketing Association's Advanced Research Techniques Forum, Monterey, CA (June).

Ehrenberg, Andrew (1988). *Repeat Buying: Facts, Theory and Applications* (2nd edition). Griffin, London.

Green, P.E. and Krieger, A.M. (1993) Conjoint analysis with product positioning applications. In J. Eliashberg and G.J. Lilien (eds), *Marketing*, Handbooks in Operations Research and Management Science, Vol. 5. North-Holland, Amsterdam, pp. 467–516.

Gunaratne, K.A. (2001) Chewing gum marketing – a conjoint analysis example. UNITEC Institute of Technology.

Hunt, J.D. (1992) Report on EMME/2 logit mode split model development for the City of Calgary. City of Calgary, Canada.

Hunt J.D., Abraham, J.E. and Patterson, D.M. (1995) Computer generated conjoint analysis surveys for investigating citizen preferences. In R. Wyatt and H. Hossain (eds), *Proceedings of the Fourth International Conference on Computers in Urban Planning and Urban Management*, Volume 2. Department of Geography and Environmental Studies, University of Melbourne, pp. 13–25.

Krusakal, J.B. (1965). Analysis of factorial experiments by estimating monotone transformations of the data. *Journal of the Royal Statistical Society*, Series B, 27, 251–63.

MacLean, Scott and Croft, Vivienne (1993) Pricing and Packaging the 1993 Indy Grand Prix – An application of conjoint measurement. In *Taking Focus on Consumers, Services and Brands*. Market Research Society of Australia and Australian Marketing Institute.

Orme, B. and Huber, J. (2000). Improving the value of conjoint simulations. *Marketing Research*, 12 (Winter), 12–21.

Srinivasan, V. and Shocker, A.D. (1973) Linear programming techniques for multi-dimensional analysis of preferences. *Psychometrika*, 38, 337–69.

Part 4
More Advanced Techniques

8

Path Analysis and Structural Equation Modelling

8.1   What is path analysis?

Path analysis is an extension of multiple regression analysis. Its purpose is to examine the magnitude and significance of causal connections between sets of variables. The causal connections are hypothesized by the researcher. While in multiple regression analysis all independent variables are assumed to affect the dependent variable directly, path analysis assumes a more complex set of relationships.

For example, assume that customer satisfaction is affected by product quality, service quality, and company image. In multiple regression analysis, our conceptual model will look like the one in Exhibit 8.1(a). In path analysis, the model can be more complex. We may hypothesize that: product quality directly affects customer satisfaction; service quality directly affects customer satisfaction; both service quality and product quality affect company image; and company image affects customer satisfaction. These hypothesized relationships are shown in Exhibit 8.1(b). The objective of path analysis is to determine: how much service quality contributes directly to customer satisfaction; how much product quality contributes directly to customer satisfaction; how much company image contributes directly to customer satisfaction; how much service quality contributes indirectly (through company image) to customer satisfaction; how much product quality contributes indirectly (through company image) to customer satisfaction. The directional lines are known as *paths*. Path analysis is no more than a series of multiple regression equations set up as follows:

$$CS = a_1 + b_{11}SQ + b_{12}PQ + b_{13}CI + e_1,$$
$$CI = a_2 + b_{21}SQ + b_{22}PQ + e_2.$$

Path coefficients are no more than standardized regression coefficients (bs) derived from the above equations. The intercepts are ignored. If estimates of error (e_1, e_2) are required, they are calculated as $1 - R^2$ (not $1 -$ adjusted R^2). Unlike in straightforward multiple regression, in path analysis the same variable can be both a dependent and an independent variable (company image in our example). Paths for which regression coefficients are not significant will be dropped from the final model. A more complicated example of a path analysis diagram is given in Exhibit 1 of the

EXHIBIT 8.1
_____

Comparison of conceptual models in multiple regression analysis and path analysis

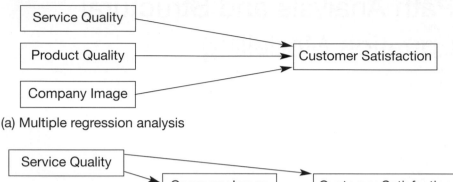

(a) Multiple regression analysis

(b) Path analysis

reading in Section 8.3.

Despite the technique's apparent sophistication, path analysis cannot really confirm the model specified by the researcher. It can no more (automatically) identify spurious relationships than simple correlations. What it does do is give a quantitative expression to the model created by the analyst.

8.2   What is structural equation modelling?

Path analysis is a subset of a more elaborate technique known as structural equation modelling (SEM). The most important difference between path analysis and SEM is that SEM allows for measured and latent variables. A *measured variable* is a variable that can be observed directly and is measurable. Latent variables are those that are not directly measured but are implied by the covariances among two or more measured variables. They are also known as factors (as in factor analysis), constructs or unobserved variables. In effect, path analysis is a simple extension of multiple regression analysis on measured variables, while SEM is a more complicated extension of multiple regression on factors extracted from measured variables.

Structural equation models consist of a measurement model and a structural model. The measurement model determines the relationships between measured variables and latent variables (factors) while the structural model deals with the relationships between latent variables. Latent variables are free of random error, because error has been estimated and removed, leaving only a common variance. An example of a structural equation model is given in Exhibit 2 of the reading in Section 8.3.

Developing and testing a structural equation model

The objective of building a structural equation model is to find a model that fits the data well enough to serve as a parsimonious representation of reality. It involves the following steps:

- Model specification
- Model identification
- Model estimation
- Model fit testing
- Model modification

These steps are briefly explained below.

Model specification

In this step, the researcher specifies causal paths among latent variables by specifying which parameters are fixed and which ones are free. *Fixed parameters* are not estimated from the data; they are usually given a value of zero, indicating no relationship between variables, and no path is drawn. (Fixed parameters, when not assigned a value of 0, are labelled numerically.) *Free parameters* are estimated from the observed data. The decisions as to which parameters are free and which are fixed are based on *a priori* hypotheses of the researcher.

Model identification

In the next step, the researcher needs to identify whether there is sufficient information to estimate the unknown parameters in the SEM. In the model, known parameters (correlations and covariances)) among latent variables must equal or exceed the number of unknown parameters (path, correlations between exogenous variables and correlated disturbance between variables). When a model contains more unknown parameters than known ones, it is *under-identified*, and when a model contains more known parameters than unknown ones, it is *over-identified*; and when both sets are equal, the model is *just identified*. The parameters of under-identified models cannot be estimated.

The number of known parameters is given by $[n(n - 1)]/2$, where n is the total number of variables in the model. For a model to be just identified or over-identified, we need to ensure that the number of the paths, correlations and disturbances between latent variables does not exceed the known parameters derived as above.

Model estimation

To generate an estimated population covariance matrix, $\Sigma(\theta)$, from the model, the researcher needs to input the start values of the free parameters. Start values can be based on prior knowledge, generated by computer programs used to build SEMs, or derived from multiple regression. The purpose of estimation is to generate a $\Sigma(\theta)$ matrix that minimizes the difference between the population covariance matrix $\Sigma(\theta)$ and the observed covariance matrix $\mathbf{S}$. The minimization function is of the form:

$$F = [\mathbf{s} - \boldsymbol{\sigma}(\boldsymbol{\theta})]'\mathbf{W}[\mathbf{s} - \boldsymbol{\sigma}(\boldsymbol{\theta})],$$

where $\mathbf{s}$ is a vector containing the variances and covariances of the observed variables, $\boldsymbol{\sigma}(\boldsymbol{\theta})$ is a vector containing corresponding variances and covariances as predicted by the model, and $\mathbf{W}$ is a weight matrix. The estimation method determines the weight matrix $\mathbf{W}$. $\mathbf{W}$ is chosen to minimize F, and F $(n - 1)$ gives the fitting function, in most cases a χ^2-distributed statistic. Generalized least squares (GLS), maximum likelihood (ML) and asymptotically distribution-free (ADF) are commonly used estimation methods. GLS is given by

$$F_{GLS} = \frac{1}{2} \, tr[([S - \Sigma(\theta)]W^{-1})^2],$$

where tr is the trace operator, which takes the sum of the elements on the main diagonal of the matrix, and W^{-1} is an optimal weight matrix, which must be selected by the researcher (the most common choice is S^{-1}).

ML is given by

$$F_{ML} = \log|\Sigma| - \log|S| + tr(S\Sigma^{-1}) - p,$$

where, $\Sigma^{-1} = W$, and p is the number of measured variables.

As for ADF, we have

$$F_{ADF} = [S - \sigma(\theta)]' W^{-1} [S - \sigma(\theta)],$$

where W contains elements that take into account kurtosis.

ML and GLS are used for normally distributed data when factors and errors are independent whereas ADF is used for non-normally distributed data. But because ADF is shown only to work well with sample sizes above 2500 (not very common in marketing research), other methods may be considered for non-normal data. No matter which procedure we choose, our aim is to obtain a fitting function that is close to 0. A fitting function score of 0 implies that the model's estimated covariance matrix and the original sample covariance matrix are equal.

Model fit testing

Although we would like the fitting function to be as close to 0 as possible, a ratio between χ^2 and degrees of freedom of 2 or less is generally considered to be a good fit. There are also a number of fitting functions not based on χ^2.

Model modification

If the variance–covariance matrix estimated by the model does not adequately reproduce the sample variance–covariance matrix, the model is revised and retested, by adding and eliminating pathways specified in the original model. The commonly used tests are the Lagrange multiplier index (LM) and the Wald test, both of which identify the change in χ^2 value when models are revised. The LM tests to see whether the addition of free parameters increases the model fit, while the Wald test asks whether deletion of free parameters increases the model fit.

The final step (after creating a model with an acceptable fit) is the estimation of free parameters. All free parameters are tested for significance. After the individual relationships within the model are assessed, parameter estimates are standardized for final model presentation. Standardized parameters are then interpreted with reference to other parameters in the model and the relative strength of pathways within the model can be compared.

Structural equation modelling is complex and is generally carried out using well-known computer programs such as LISREL, Amos, EQS or PISTE.

Limitations of SEM

As with path analysis, the directions of arrows in a structural equation model represent the researcher's hypotheses of causality within a system. The researcher's choice of manifest and latent variables as well as the paths represented will restrict the structural equation model's ability to reproduce the sample variance and covariance patterns that have been observed in reality. Consequently, it is possible – and quite common – to find two or more models that fit the data equally well. There is no way of telling which one is the 'true' representation of reality.

In spite of its limitations, SEM remains a popular approach to analysing complex causal data in marketing research, especially in the areas of customer satisfaction, value and loyalty. Although the model itself cannot distinguish between indirect and direct relationships and variables, once the researcher specifies the model, SEM can analyse relationships between latent variables and quantify them.

8.3   Application of path analysis and structural equation modelling

In the following article Terry Grapentine (2000) compares path analysis and structural equation models using an example from customer satisfaction research. Grapentine also discusses the relative strengths and weaknesses of each approach and makes recommendations to the practitioner. (To simplify the discussion, the term 'path analysis' will always refer to regression-based path analysis, and SEM to LISREL-type structural equation modelling incorporating maximum likelihood estimation procedures.)

Do the relative merits of path analysis and structural equation modeling outweigh their limitations?

Marketing research departments are often interested in quantifying the relative importance of factors affecting issues such as brand equity, brand loyalty, and customer satisfaction. Two widely used methods in this regard are path analysis (incorporating ordinary least-squares regression) and structural equation modeling (SEM) (typically incorporating maximum likelihood estimation). The latter is considered to be the more 'advanced' of the two. But is it?

Overview

Both methods [path analysis and structural equation modeling] are forms of causal modeling that examine relationships between and among one or more dependent variables and two or more predictor or independent variables. Examples of dependent variables might be measures of customer satisfaction, loyalty toward a brand or company, brand equity, or perceived value. Independent variables frequently focus on issues relating to brand or company performance such as perceived product quality, service quality, and price competitiveness. Both methods help decision makers understand the relationships between the independent and dependent variables.

Neither path analysis nor SEM are methods for discovering causative relationships. Rather, they are a means by which theoretical relationships can be tested. In applied research, one often sees path analysis employed to test relatively simple relationships such as displayed in Exhibit 1. In contrast, SEM is sometimes referred to as latent-variable analysis because these models establish the relationships between 'unobserved' variables – a topic that will be discussed in more detail shortly. In the practical world, SEM is sometimes employed to examine more complicated relationships that cannot be handled by path analysis (i.e., models incorporating reciprocal causation).

Path Analysis Example

Exhibit 1 presents a path analysis example for a consumer electronics product. All variables are represented by rectangles, which denote 'observed' variables. These variables are measured by asking respondents questions

EXHIBIT 1

Consumer electronic product path analysis

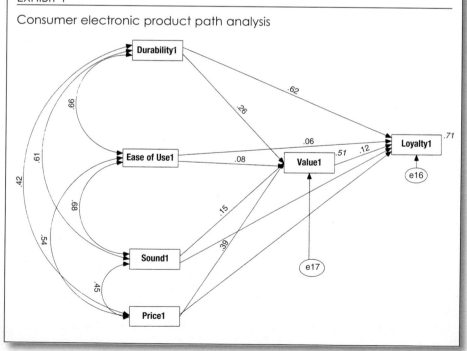

related to the construct represented by the rectangle. Such questions typically have respondents rate a product on a series of attributes using some kind of scale where higher scale numbers denote better-perceived performance.

In most cases, the variables denoted by the rectangles are *summated scales* where two or more attributes measuring a common underlying construct are summed and divided by the number of items. For example, if four attribute statements measured the Durability construct – for each respondent – one simply adds the four items together and divides by four.

Summated scales produce two benefits in models such as these. First, they help manage multicollinearity's effects on the estimation of regression coefficients and second, they help focus management's attention on more fundamental dimensions of product/company performance, of which the individual attributes are indicators.

The straight arrows from one observed variable to another denote the functional relationships between the variables, which are specified by the marketing researcher prior to analysis. In Exhibit 1, Durability1 is functionally related to Value1. If the perceived performance of Durability1 increases or decreases, Value1 increases or decreases as well. The value construct, in turn, is functionally related to brand loyalty (Loyalty1).

More formally, variables that only have straight arrows pointing from them to another variable – for example, Durability1, Ease of Use1, Sound1, and Price1 – are called exogenous variables. This is because their values are not determined by any other variables in the model (i.e., no arrows from other constructs point to them). In contrast, Loyalty1 is an endogenous variable because its value is determined by other constructs in the model (e.g., Durability1, Ease of Use1, Sound1, Price1, and Value1). The various arrows pointing from these constructs to Loyalty1 represent these functional relationships. Value1 is called a mediating variable because it serves to mediate the effect selected exogenous variables have on the endogenous variable. (See the sidebar *Testing for Mediation Effects* on p. 171.)

The numbers near the straight arrows are the standardized regression co-efficients obtained when an endogenous variable is regressed on the set of exogenous variables to which it is functionally related. (The regression coefficients are standardized as a result of standardizing the independent and dependent variables so each variable has a mean of zero and a stan-dard deviation of one.) For example, in Exhibit 1, regressing Value1 on the set of exogenous variables that point to it produces the following stan-dardized regression coefficients: Durability1 = .26, Ease of Use1 = .08, Sound1 = .15, and Price1 = .39.

The curved double-headed arrows linking the exogenous variables to each other represent the fact that these exogenous variables are corre-lated. (Unless one is conducting experimental design research, this situa-tion is a fact of life in survey research.) The numbers by the curved arrows are the correlation coefficients between each of the variables. For exam-ple, in Exhibit 1, the correlation between Durability1 and Ease of Use1 is .66.

Effects of Exogenous Variables

In Exhibit 1, each exogenous variable affects Loyalty1 directly and indir-ectly. The arrows linking the exogenous variables to Value1, and then link-ing Value1 to Loyalty1, suggest the indirect linkage. In this role, Value1 is a mediating variable. To calculate the total effect that an exogenous vari-able has on Loyalty1, one simply adds the direct and indirect effects together. In this example, the direct effect of Durability1 on Loyalty1 is .62. The indirect effect is calculated by multiplying the coefficient represented by the arrow from Durability1 to Value1 (.26) by the coefficient represented by the arrow from Value1 to Loyalty1 (.12), or .26 × .12 = .03. The total effect of Durability1 on Loyalty1, therefore, is .62 + .03 = .65.

Measurement without Error

A critical assumption underlying the use of regression analysis in calculat-ing these coefficients is the endogenous and exogenous variables are measured without random error. This assumption usually isn't true. Measurement error creeps into questionnaires from many sources, which generally can be classified into two categories – systematic and random error. Systematic error refers to measurement error that is directional or constant in nature. Gilbert Churchill describes this type of error, which results in biased measures, in his 1999 book *Marketing Research: Methodological Foundations* as: 'Systematic error is also known as constant error, because it affects the measurement in a constant way. An example would be the measurement of a man's height with a poorly calibrated wooden yardstick.'

A marketing research example could be a study that purports to measure the quality of waiters or waitresses in a restaurant. The domain of this construct might include issues such as their courtesy, cleanliness, product knowledge, and friendliness. If in creating this construct measure (through a summated scale, for example) we omit a measure of product knowledge, we inadvertently create a biased sampling of the domain of the concept, which results in systematic error in our measure.

Random measurement error, by implication, is non-systematic error. An example of random error might be the fact that respondents are not always consistent in their answers to survey questions from one point in time to another (assuming of course there is no change in the attitude or perception being measured).

As far as the computer is concerned, the data that are input into a regression analysis computer program are error free. This assumption is often false and, consequently, the measurement error 'effect' is contained in the error term of the equation:

$$Y = a + \beta x_1 + \beta x_2 + \beta x_3 + \ldots + \beta x_n + \epsilon.$$

Also contained in the error term are the effects of other variables important in predicting Y, but are missing in the equation.

Structural Equation Modeling Examples

Refer to Exhibit 2, which gives the SEM counterpart for the path analysis model of Exhibit 1. This SEM model is calculated using the Amos software package, Version 4.0. Clearly, there are many visual differences between path analyses and SEM.

EXHIBIT 2

Consumer electronic product maximum likelihood estimation

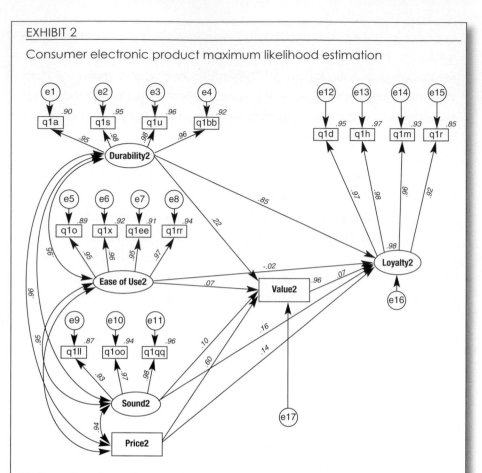

First, and perhaps foremost, SEM attempts to account for *random* measurement error. That is what all the small circles with the letter 'e' in the middle of them suggest. Consequently, the standardized coefficients in a structural equation model possess more reliable estimates of how an exogenous variable affects an endogenous variable than what is produced with path analysis.

Coefficient Estimation in Path Analysis and SEM

In calculating path analysis coefficients, ordinary least-squares regression analysis minimizes the squared distance between the estimated Y variable (the dependent or exogenous variable) and the actual value of Y. In contrast, SEM typically incorporates the covariance matrix of the independent and dependent variables. It uses a maximum likelihood estimation procedure to derive the 'most likely' coefficient values, given the actual covariance matrix. (The covariance between any two variables measures the extent that a change in one variable is associated with a change in another variable. Covariance is not the same as correlation, but they are related. The correlation between two variables is equal to the covariance of the two variables divided by the product of the variables' standard deviations.)

Observed vs. Latent Variables

One obvious difference between the path model and the structural equation model is that all path analysis model variables are contained in rectangular boxes, and most of the structural equation model variables are represented by ovals. Each oval has one or more small boxes linked to it with arrows pointing from the ovals to the small boxes.

In Exhibit 2, q1a, q1s, q1u, and q1bb are four attribute statements that serve as measures or indicators of the unobserved variable, Durability2. In this structural equation model, Durability2 is an unobserved or latent variable. (In Exhibit 1, Durability1 is an observed variable because it's a summated scale comprised of these same four variables.) The arrows pointing from Durability2 to the q's denote the variance in the observed variables (the q's) is caused by an underlying construct that we cannot directly observe, but which the researcher believes is important in predicting variables Value2 and Loyalty2. The 'e's', on top of the q's, represent the random error associated with Durability2's ability to completely explain the variance in the q's. SEM takes this measurement error into account when calculating the model's coefficients – path analysis doesn't.

The numbers beside the arrows pointing from Durability2 to the q's can be considered as standardized regression coefficients. The larger the number, the more a q variable can be considered a good indicator of the latent variable. The numbers beside the arrows – from the e's to the q's – represent the amount of variance in the q's that can be explained by the unobserved variable, Durability2. The larger the number, the more the unobserved variable can explain the variance in the q's.

As with path analysis, the structural equation model also allows us to derive the total effect that each of the exogenous variables has on Loyalty2 (the direct and indirect effects combined) by using the same procedures discussed earlier for path analysis. In the path analysis model, the total effect of Durability1 on Loyalty1 is .65. In the corresponding structural equation model, the total effect of Durability2 on Loyalty2 is .65 + [.22 × .07] = .67. These results are pretty close. And keep in mind, both models use the same data set. They differ primarily in how each goes about estimating the model's coefficients. Moreover, SEM uses a confirmatory factor analysis approach to measure unobserved variables, whereas the path model only uses observed variables.

Are Model Results Different?

The rank order of importance of the total effects of the exogenous variables on Loyalty is comparable between the path and SEM models but not identical (see Exhibit 3). The most important variable in the path analysis model is also the most important variable in the structural equation model. The other variables' rank orders are similar but not identical. Management decisions based on either model would likely be comparable as well. Yet, the academic community says that SEM is the better of the two methods for the following reasons:

- The coefficient estimates are more valid because the estimation procedures take random measurement error into account.
- r^2 is higher in the structural equation models.
- Structural equation models with latent variables use a confirmatory factor analysis technique that provides information on which observed variables are the best indicators of the unobserved variables.

And yet, there may be a problem. When SEM takes random measurement error into account, multicollinearity can increase, and the parameter estimates become less stable. (This is a matter of degree, however, depending upon how reliable the summated scale scores are in the path analysis models.) Moreover, the percentage of the variance explained in the dependent variable, as measured by adjusted r^2, can increase to levels that lack credibility.

Testing for Mediation Effects

The mediator variable in Exhibits 1 and 2 is Value because the exogenous variables can affect brand loyalty both directly and indirectly through the value–loyalty linkage.

Examination of the mediation issue is important for two reasons. First, from a theoretical perspective, if a construct such as Value mediates the effect that exogenous variables have on a dependent variable such as Loyalty, the model should incorporate these functional relationships. Second, in creating models with mediator variables, the practitioner must answer the following question: 'Is the mediator variable in my model statistically significant?'[1]

Authors Ruben M. Barron and David Kenny developed a procedure for testing the statistical significance of mediator variables. (For more explanation, see their 1986 *Journal of Marketing Research* article listed in the Additional Readings section below.) Their approach utilizes ordinary least-squares multiple regression, which is available in most statistical software packages. The procedure involves the following three steps, using Exhibit 1 as an example:

Step 1: Regress the mediator variable, Value1, on the explanatory variables, Durability1, Ease of Use1, Sound1, and Price1.

Step 2: Regress the dependent variable, Loyalty1, on the same set of explanatory variables in Step 1.

Step 3: Regress Loyalty1 on the explanatory variables of Step 1 and the mediator variable, Value1.

According to Sanjeev Agarwal and R. Kenneth Teas, from their 1997 Iowa State University working paper, 'the procedure produces evidence of mediation when (a) explanatory variables are statistically significant in the estimates of Step 1; (b) the explanatory variables are statistically significant in the estimates of Step 2; and (c) the mediator variable is statistically significant in the estimates of Step 3'.

The table below shows the mediator variable, Value1, is statistically significant. Additionally, as discussed below, there are three mediation outcomes – no mediation, partial mediation, and full mediation.

Testing for mediation effects (data drawn from Exhibit 1)

Standardized regression coefficient

(given if significant at .05 or less)

	Dependent variable			
	Step 1:	Step 2:	Step 3:	
Variable	Value1	Loyalty1	Loyalty1	Interpretation
Durability1	.26	.65	.62	Partial mediation
Ease of Use1	.08	.07	.06	Partial mediation
Sound1	.15	.14	.12	Partial mediation
Price1	.39	.08	ns	Full mediation
Value1			.12	Significance is required for any kind of mediation

No mediation exists if the regression coefficient for an exogenous variable is insignificant in Step 1 or if the regression coefficient for Value1 in Step 3 is insignificant (which it's not). Partial mediation exists if the regression coefficient for an exogenous variable is significant in Steps 1 and 3 (as is the case with Durability1, Ease of Use1, and Sound1).[2] Full mediation exists if the regression coefficient for an exogenous variable such as Price1 is significant in Step 1 but not in Step 3. (Although Step 2 is not referenced in the above table, conducting Step 2 provides a more comprehensive test for mediation effects. For example, evidence for mediation is increased if the standardized beta coefficients of the independent variables in Step 2 are significantly greater than comparable values in Step 3, which is not the case in this particular example.)

In summary, there are three possible mediation outcomes for each exogenous variable – no mediation, partial mediation, and full mediation. Exogenous variables that have partial mediation (i.e., a direct and indirect linkage to a dependent variable) are generally more important predictors of a dependent variable such as Loyalty than similar variables that have full mediation. Consequently, if moderator effects are not examined appropriately, management can be misled as to the relative importance of different factors affecting a construct such as Loyalty.

[1]In reality, the mental processes by which consumers make decisions are too complex for any model to specify all mediator variables – or all exogenous variables, for that matter, No matter how good a model appears to be it will always omit constructs, which represent attitudes or perceptions of consumers that affect consumer behaviour. We can only hope that we include in our moels those mediator variables that theory suggests that are important in our attempt to understand in greater detail how consumers make decisions.

[2]A situation where an exogenous variable is significant in Step 3 and insignificant in Step 2 is highly unlikely. The converse is not true and would represent a case of full mediation if the exogenous variable were significant in Step 1, which is the case with Price1.

Problems with Measuring Consumer Attitudes and Perceptions

Why this occurs relates to a fundamental problem that often happens when measuring consumer attitudes and perceptions in survey research. By their very nature, measures of these attitudes and perceptions are correlated. Consider the following:

- To some extent a halo effect operates when consumers are evaluating products through rating scales on a quantitative survey instrument. For example, if a respondent is rating two products and one product is preferred over the other, the more preferred product will generally receive more favorable ratings than the less preferred product, regardless of the attribute.
- This halo effect can be exacerbated when respondents do not carefully consider how a product performs on a specific attribute and, perhaps because of interview length, become lazy in assigning numerical ratings to product attributes.
- Finally, random error in data sets attenuates the strength of the relationship among functionally related variables. By controlling for random error, structural equation models will generally produce higher r^2s than path models that don't take this source of error into account, everything else held constant.

EXHIBIT 3

Total effect of exogenous variables on loyalty: consumer electronic product

Variable	Path Analysis		SEM	
	Coefficient	Rank	Coefficient	Rank
Durability	.65	1	.67	1
Ease of Use	.07	3–4	0	4
Sound	.14	2	.17	2–3
Price	.09	3–4	.18	2–3
Adjusted r^2	.71		.98	

Examine the correlations among the independent or exogenous variables between the path model and the structural equation model. In nearly all instances, the correlations are higher in the structural equation model than in the path model, indicating a higher level of multicollinearity. For example, the correlation between Ease of Use and Durability in the path model is .66. The correlation between the same two constructs in the corresponding structural equation model is .95, which is nearly perfectly collinear! The effects of multicollinearity are as follows:

- Model coefficients may be far from the true, unknown parameters of which they are estimates. In extreme cases, individual coefficients may take on the wrong sign – that is, the coefficients take on negative values when theory or common sense suggests a positive relationship exists between the independent and dependent variables.
- The standard errors of the coefficients are likely to be large and, consequently, a coefficient can be insignificant (i.e., not significantly different from zero), even though the true value of beta is not zero.

Further evidence of the effects of multicollinearity can be shown in the inflated standard errors of the models' coefficients in the structural equation model vs. the path analysis model by using a bootstrapping technique.[1]

According to James Arbuckle and Werner Wothke, in their book *Amos 4.0 User's Guide*, bootstrapping 'is a versatile method for evaluating the empirical sampling distribution of parameter estimates. In particular, bootstrapping can be used to obtain empirical standard error estimates of the model parameters.'

Bootstrapping is a computationally intensive simulation approach. It treats the original sample as a stand-in for the data population and draws many bootstrap samples from it by sampling with replacement. Each bootstrap sample is used to obtain a new set of parameter estimates. The distribution of each parameter estimate across bootstrap samples is used to obtain empirical estimates of its estimation bias, standard error, and confidence intervals.

Exhibit 4 shows the relatively larger standard errors in the structural equation model compared to the path model. The standard errors for the structural equation models range from 1.13 to 1.71 times as large as for the path analysis model.[2] What does this mean to the research practitioner? Consider the following:

- Path analysis (or for that matter, simple or multiple regression equations) may disguise multicollinearity's effects (We, of course, do not include situations where regression is used in controlled experiments where the independent variables are orthogonal to each other.) This is because regression analysis assumes the independent variables are measured without random error, which in nearly all instances in applied survey research is an incorrect assumption.

---

[1]Other factors that can cause inflation in these standard error estimates are (1) more parameters estimated with the same sample size, resulting in lower efficiency of each estimate and (2) non-linear item to factor relations, resulting in larger bootstrapped standard errors.

[2]These as well as other findings are not unique to the data set on which this article is based. Earlier drafts on this artile demonstrated similar findings in two additional studies from two different product categories. In some cases, the standard errors for the structural equation models were over 3 times as large as for the corresponding variables in the path models.

• SEM takes into account random measurement error, and as a result, more reliably exposes multicollinearity's effects. This as well as other factors (refer to note 1) unfortunately can cause an increase in most, if not all, of the standard error estimates of the model's coefficients. Increased standard errors mean less stable coefficients.

EXHIBIT 4

Bootstrapped estimates: path analysis vs. structural equation models

Model	Dependent variable	Independent variables	(1) Path analysis	(2) Structural equation models	(2) ÷ (1)
Consumer Electronics	Loyalty	Durability	.035	.049	1.40
		Ease of Use	.036	.056	1.56
		Sound	.038	.065	1.71
		Price	.031	.035	1.13
		Value	.039	.045	1.15

What does this mean for the marketing research analyst? My answer is that research practitioners should more often consider using alternative research methods that control for multicollinearity as opposed to always collecting and analyzing cross-sectional data and then blindly using path analysis or SEM to analyze it, which is how most survey research is conducted today. Examples of alternative methods are conjoint analysis and research designs incorporating control and test groups where stimuli are manipulated in ways to eliminate the effect multicollinearity or other factors may have on research findings.

Admittedly, experimental designs are not always feasible in a practical setting. If one, therefore, is limited to using cross-sectional data and path analysis or SEM, one needs to insure the model is grounded in a solid theoretical framework. (See my article entitled 'Practical Theory', from the

Summer 1998 issue of *Marketing Research: A Magazine of Management and Applications*, for further explanation.)

Additionally, if you know a priori that two highly correlated variables are critically important in affecting brand loyalty, you might consider an altogether different research design to assess which is more 'important' to customers. For example, you could use a series of direct questions to determine where a company should seek to improve performance to maximize customer value. (Although in practice, this too can be a tricky task and raises the topic of what one means by the term *importance*.)

Conclusions

Researchers use regression-based path analysis and SEM to understand how consumer attitudes and perceptions affect behavior. Graphically, path and structural equation models look similar. Moreover, both models produce comparable, although not identical results. It the models are not too complex, it's quite likely that path analysis and structural equation models will identify the same 'most important' and 'least important' variables, although the rank order of variable importance may not be completely the same.

Path analysis, however, assumes that survey measures are made without random measurement error. As a result, this method can partially disguise multicollinearity's effects. Although SEM explicitly takes random measurement error into account, the results can produce less stability in the model's estimated coefficients due to higher coefficient standard errors caused in part by multicollinearity. Consequently, the research practitioner may be well advised to consider alternative research designs that control for multicollinearity as well as other factors that can affect the reliability and validity of research findings.

(Terry Grapentine is the principal of Ankeny, Iowa-based Grapentine Company Inc. He is also a member of Marketing Research *magazine's editorial review board. The author thanks Dr. Werner Wothke, principal of Chicago-based SmallWaters Corp. and creator of Amos, for helpful comments on early drafts of this article.)*

Additional Reading

Agarwal, Sanjeev and Teas, R. Kenneth (1997) Quality signals and perceptions of quality, sacrifice, value, and willingness-to-buy: An examination of cross-national applicability. Working Paper No 37–16, Iowa State University.

Arbuckle, James L. and Wothke, Werner (1999) *Amos 4.0 User's Guide*. SmallWaters Corporation, Chicago.

Baron, Ruben M. and Kenny, David A. (1986) The moderator–mediator variable distinction in social psychological research: Conceptual, strategic, and statistical considerations. *Journal of Marketing Research*, **19** (May), 229–39.

Churchill, Gilbert A., Jr (1999) *Marketing Research: Methodological Foundations* (7th Edition). Dryden Press, New York.

Grapentine, Terry (1995) Dimensions of an attribute. *Marketing Research: A Magazine of Management and Applications*, **7** (Summer), 19–27.

Grapentine, Terry (1998) Practical theory. *Marketing Research: A Magazine of Management and Applications*, **6** (Summer), 5–12.

Hoyle, Rick H. (ed.) (1995) *Structural Equation Modeling: Concepts, Issues, and Applications*. Sage, Thousand Oaks, CA.

Schumacker, Randall E. and Lomax, Richard G. (1996) *A Beginner's Guide to Structural Equation Modeling*. Lawrence Erlbaum Associates, Mahwah, NJ.

Bibliography

Further reading

For a relatively simple introduction to structural equation models consult

Schumacker, Randall E. and Lomax, Richard G. (1996) *A Beginner's Guide to Structural Equation Modeling*. Lawrence Erlbaum Associates, Mahwah, NJ.

Reference

Grapentine, Terry (2000) Do the relative merits of path analysis and structural equation modelling outweigh their limitations? *Marketing Research*, **12**(2), 13–21.

9

Data Mining Techniques

9.1   What is data mining?

Like any other discipline, marketing is concerned with finding predictable relationships among variables. What are the characteristics of loyal consumers? What attitudes and behaviours lead to a purchase? Why do people switch brands? What attributes and characteristics of a customer contribute to his or her creditworthiness?

Human behaviour being complex, such questions are not easily answered. For instance, customer loyalty may involve a number of factors such as a person's age, gender, place of residence, income level, marital status, availability of alternatives, and past purchase patterns, to name a few. Even when we know the factors that influence purchase behaviour, there is still the problem of knowing the combination of characteristics that would best predict what we are interested in – in this example, customer loyalty. The major problem here is that thousands or even millions of combinations of predictor variables are possible. It is impossible to sift through all possible combinations of relationships except by mechanical means.

Data mining is the mechanical search for patterns and relationships in data. Consider these examples:

1. A large bank has considerable information on its customers. A customer asks for a large loan. From the characteristics of the customer – age, income, place of residence, marital status, number of children, net worth, date of inception of account, average monthly balance, etc. – can the bank calculate beforehand how risky it is to lend money to this customer?
2. A supermarket chain wants to assess what non-baby products it should stock in the baby products aisle, so it can increase the sale of unrelated items when a customer comes to buy baby products. This means understanding the patterns of purchase by thousands of customers of thousands of products.
3. A hotel chain would like to understand the patterns of guest registration and behaviour so it can offer discounted rates at slow periods to attract additional sales without greatly reducing its revenue through discounting.
4. A telephone company is interested in managing its customer relationships based on individual customer characteristics.

All the above instances involve extensive analyses of databases and complex computations. Techniques of data mining can be applied in all such contexts. From a broad perspective, data mining involves all of the following:

- data collection (data warehousing, web crawling);
- data cleaning (dealing with outliers, errors);
- feature extraction (identifying attributes of interest);
- pattern discovery and pattern extraction;
- data visualization;
- results evaluation.

However, from a more focused perspective, the term 'data mining' is generally applied to pattern discovery and pattern extraction.

Conceptually, data mining applications in marketing fall into two broad categories – grouping of variables and identifying functional relationships among variables. While the basic multivariate techniques discussed in this book thus far cover these two aspects, data mining techniques are concerned with achieving these objectives with potentially large databases with a large number of variables. Condensing a large number of variables into potentially meaningful groups and finding relationships among a large number of variables – often without any underlying hypotheses – characterize these techniques. In that they resemble exploratory data analysis.

Although data mining may be the only available way to deal with the analysis of complex patterns, it is best to consider it as an exploratory technique to generate hypotheses rather than as a confirmatory technique that leads to conclusions. As Bonferroni's theorem warns us, if there are too many possible conclusions to draw, some will be true for purely statistical reasons, with no physical validity.

9.2 Data mining models

Data mining uses a variety of analytic tools to uncover patterns in data. Although there are many such tools, the following are the ones most frequently used:

- Data visualization
- Association rules
- Decision trees
- Case-based reasoning
- Neural networks
- Genetic algorithms.

Visualization

Data visualization takes advantage of the capacity of human beings to recognize and distinguish patterns of observable characteristics. Visualization is particularly effective for exploring and condensing large amounts of messy data into compact understandable pictures. These techniques range from exploratory techniques such as simple histograms, box plots, scatter diagrams, and link analysis networks to more complex techniques such as rotating multicoloured three-dimensional surface plots in three dimensions.

Association rules

Association rules state the relationships between the attributes of a group of individuals and one or more aspects of their behaviour. The purpose of these rules is to enable predictions about the behaviour of other individuals who are not in the group but possess the same attributes. Association rules are stated in dichotomous terms such as good credit risk vs. bad credit risk, buyer vs. non-buyer. These rules assign probability-like numbers to actions.

Association rules are of the form $\{X_1, X_2, ..., X_n\} \Rightarrow Y$: if we find all of $X_1, X_2, ...,$ X_n, then we have a high probability of finding Y. As an example of an association rule, suppose a mail-order institution is interested in cross-selling a personal digital assistant to those who have just ordered several electronic items. Promoting a PDA to unlikely customers may antagonize them while wasting the salespeople's time. Therefore, the company would like to restrict the offer to customers who have a high probability of buying a PDA. To accomplish this the company can analyse purchase records with association rule methods. Such analysis may indicate that customers who bought a CD player and a wireless telephone on one call were much more likely to buy a PDA on a subsequent call than customers who ordered tape recorders or calculators. Consequently, when the association rules are incorporated in the company's order entry system and the system identifies that the customer on the phone recently ordered a CD player and a wireless telephone, it prompts the sales person to make the offer on a PDA. On the other hand, if the system finds that the caller bought a tape recorder or a calculator on the last order, the built-in decision rules will not prompt the salesperson to make that offer (but presumably an offer of another product which has a high probability of purchase for those who bought tape recorders or calculators).

The probability level of finding Y for us to accept this rule is known as the *confi dence* of the rule. Generally, we would search only for rules whose confidence is above a certain threshold and is significantly higher than what would be obtained if Xs are chosen at random. The purpose of the latter condition is to avoid spurious associations. For instance, a supermarket might find a rule like {chocolate, cigarette} $\Rightarrow$ newspaper, but that might only be because a lot of people buy newspapers, irrespective of what else they might buy.

Tree-based methods (decision trees)

Tree-based methods are used to sequentially partition the data set using independent variables in order to identify subgroups that contribute most to the dependent variable. The most commonly used techniques for automatic sequential splitting are chi-squared automatic interaction detectors (CHAID) and classification and regression trees (CART). Tree-based methods are good at identifying the most important variables, interactions among independent variables, and non-linear relationships. They help to identify the most important variables and eliminate the irrelevant ones. The results obtained using these methods are relatively easy for users to understand and interpret. Decision-tree algorithms are robust to outliers and erroneous data.

Consider a simple example of identifying customers of a bank who are likely to respond to a direct mail campaign; data are available on net worth, income and gender. (In practice, tree-based methods in data mining are unlikely to be used when there are only three independent variables. However, the principle of splitting is the same, whether one uses 3 or 300 independent variables.) Records show that 20% of those who were sent the direct mail responded. The objective of the analysis is to identify subgroups that will have a much higher probability of responding to the offer, based on the three independent variables. To keep things simple, let us group each of the independent variables into just three categories:[1] net worth (high, low), income (high, low) and gender (male, female). At the first level of analysis, our question is which of these three independent variables differentiates responders from non-responders. Suppose the data show that the response rates are as follows: among males 15%, and females 25%; among those with high net worth 35%, and low net worth 7%; among those with high income 30%, and low income 14%. Obviously, among the three variables, a person's net worth maximally differentiates between responders and non-responders the best. So customers are split into two groups: high net worth and low net worth. This process is repeated separately for the high net worth and the low net worth groups with the remaining two variables. When there are a number of independent variables, similar analysis is automatically performed at each stage process until a prespecified criterion is met (e.g., statistical significance or a minimum specified difference between the groups).

Exhibit 9.1 shows an example of this type of analysis. Each group can be profiled by following the tree hierarchy. Decision trees provide a defined way to develop segments in terms of a single dependent variable. However, since decision trees split the sample sequentially, they use up data rapidly and, therefore, are not suitable for use in small databases. These techniques are highly sensitive to noise in the data and they tend to *overfit* data. As a result it is important to cross-validate the findings obtained using decision-tree results.

EXHIBIT 9.1

CHAID analysis

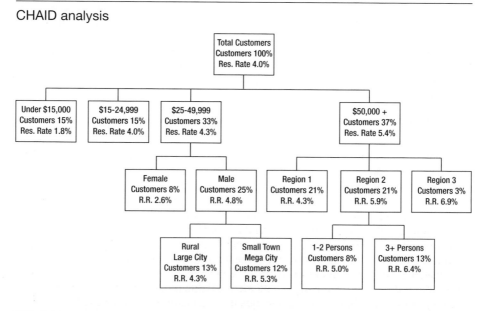

---

[1] In practice, tree-based techniques can and do split continuous independent variables into several categories.

Case-based reasoning

In case-based reasoning (CBR) systems, we compare the attributes of a new case with corresponding attributes in a collection of previously known cases. The objective here is to identify examples that provide generally positive solutions and use them to generate a template for the current case. As an example, consider a fast food chain looking into setting up new outlets. Among the specifications of such outlets, it needs to consider such matters as floor space, number of counters and whether or not to include a salad bar. It will also need to compare the attributes of the locality in which any potential new outlet is situated, such as average income, the number of teens and preteens, traffic flow, and number of commercial establishments to the corresponding attributes for all of the company's existing outlets, along with their design specifications, annual sales and profitability. The CBR system is used to identify existing locations whose attributes most closely resemble those of the proposed locations and develop design specifications for the proposed outlets. Obviously this is not meant to be a mechanical exercise since any successful site may have some features that are unique. In other words, the characteristics of successful outlets are used as a template to be modified as required and not as a mould that is inflexible.

The value of case-based reasoning systems rests on the fact that it forces the user to focus on the similarities and differences between different situations in a structured way using the attributes that define the cases. In doing so CBR translates abstract concepts into tangible attributes. CBRs are easy to understand and implement on a computer. They accommodate qualitative and quantitative variables and can deal with discontinuous variables.

On the negative side, CBRs represent *what was actually done* in the past, not necessarily *what is optimal* under similar circumstances. The solutions of the past may not necessarily be optimal under current conditions, and using them to solve current problems may simply perpetuate mistakes and suboptimal solutions of the past. Establishing and maintaining CBRs will require considerable expertise and time investment. It is not a simple task to first identify attributes that are related to specific outcomes and then to assign weights so that new situations can be matched to the most appropriate outcomes. Even more importantly, CBRs may lead to misleading conclusions when there are significant interactions among variables under consideration.

Neural networks

Neural networks are computer models that are designed to simulate human brain processes and are capable of learning from examples to find patterns in data. Although they have been around for decades, only recently have they begun to make an impact in marketing. This may be attributed to the rapidly reducing costs of computing and the emergence of better theoretical frameworks. Unlike in conventional computing, neural nets do not rely on specified methodology based on a standard set of instructions. Rather, a neural net is 'trained by example', just as a rat learns a maze or a child learns to walk.

Supervised neural network

Conceptually, a neural net is a 'black box' that produces a set of outputs on the basis of a set of inputs (Exhibit 9.2). The network is presented with a 'training set' in which both the inputs and outputs are known. Using this as an example, the network is 'trained' to model the outputs from the inputs. This is known as 'supervised training'.

EXHIBIT 9.2

The neural network black box

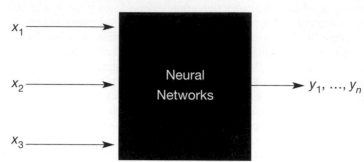

The 'black box' consists of a collection of processing units (analogous to neurons of the brain) that are connected and form a 'network'. 'Training' involves varying the weights assigned to the connections between neurons with a view to minimizing the difference between the network's outputs and the actual outputs obtained from the training set. The model thus generated is then validated by applying it to a separate 'test set' of data.

The multilayer perceptron

A common model of neural nets used in business is one known as the multilayer perceptron (MLP).[2] In the example shown in Exhibit 9.3, there are two inputs and one output, and there are three layers of neurons. The neurons in one layer are connected to every neuron in the next layer. The neuron takes the sum of its inputs and applies a function to this sum. Such a function can be either linear or non-linear (e.g., sigmoid function). A pair of input values (x_1, x_2) are presented to the input layer neurons. After being processed by the input layer, the values are passed to the connections at the hidden layer. They are then modified by applying weights $(w_1, w_2, ...)$ as they pass through this layer. The assigned starting weights are generally random but are then modified during training. The hidden layer neurons process these weighted values. The processed values are passed along the final set of connections and are modified by another set of weights $(u_1, u_2, ...)$ before reaching the output neuron. The output neuron applies an additional process to compute the value of Y_1, the network output. 'Bias units' output a fixed unit value analogous to the constant terms in the equations defining the processes carried out by the network.

As noted earlier, 'training' is essentially a process of adjusting the connection weights so as to make the network reproduce the known output values in the training set. It is the minimizing of mean square error (between the network and the training sets). Although the initial training can be time-consuming and computer-intensive when large networks are involved, once trained the weights in the network are fixed. Then it is simply a matter of calculating the output value corresponding to any pair of inputs, a relatively quick process. It is the non-linear hidden layer neurons which provide MLPs with their power in modelling data patterns. These units are frequently referred to as 'feature detectors' – they decompose the data patterns into simpler

---

[2]There are other models such as the radial basis function (RBF) which are also gaining commercial popularity.

EXHIBIT 9.3

Multilayer perceptron: an example

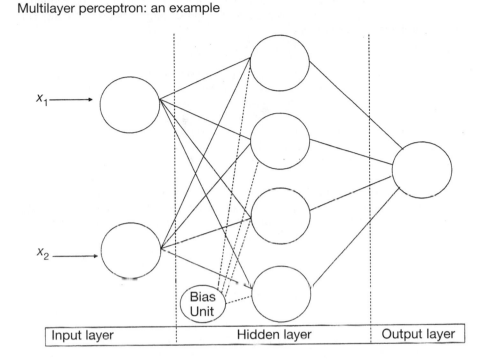

| Input layer | Hidden layer | Output layer |

features. However, when we face complicated networks, the precise role of particular hidden neurons is usually very difficult to discern. Although we used a fairly simple and straightforward illustration, neural networks cover a number of algorithms that deal with prediction, classification and clustering.

Neural nets have several advantages. They can efficiently combine information from many predictors and cope with correlated independent variables. Compared to traditional techniques such as regression and discriminant analysis, neural nets do handle non-linearities and missing data more effectively. Because of their ability to detect non-linear relationships automatically, neural nets have a significant advantage over regression-type models. Compared to standard multivariate techniques, neural net procedures and results are easier to communicate. Since these models adapt to changing input much more easily than techniques such as multiple regression analysis, they are considered to be especially appropriate in dynamic, fast-changing situations when the relationship between behaviour – (e.g., customer attrition) – and a set of predictors is subject to frequent change.

Neural nets also have several disadvantages. Building the initial neural network model can be very time-consuming since it involves extensive data cleaning, data verification, data transformation, and variable screening. Many of these procedures require specialized skills. Another main limitation of neural nets is that there is no explanation for the outcomes produced by neural nets – in most cases, it is essentially a black box technique as far as the end-user is concerned. Neural networks need to be 'trained'. Training is conceptually similar to deriving weights in a regression equation and involves reading sample data and iteratively adjusting network weights to produce a best prediction. Once such weights are assigned, the model can be applied to others to make predictions. However, training requires large amounts of data. This can be a problem in some cases.

Genetic algorithms

Genetic algorithms (GAs) are used to solve prediction and classification problems or to develop sets of decision rules similar to the rules that are inferred from decision-tree models. GAs are based on the evolutionary biological processes of selection, reproduction, mutation, and survival of the fittest. They are suitable for use with poorly understood, poorly structured problems because they aim to generate several alternative solutions simultaneously, unlike, say, a regression model which attempts to find a single best solution. GAs can also incorporate in the model any decision criterion. If, for instance, the marketer is interested in maximizing the response rate in a particular segment, this can be built into a GA model (but not into traditional multi-variate models such as logistic regression analysis). For example, a GA can explicitly model maximizing the proportion of responses in the top 20% of a direct marketing lift analysis, something logistic regression cannot do. Another feature of GAs is that they are capable of producing unexpected solutions: they may identify combinations of independent variables that may not have have been initially obvious. GAs can be used by those who may not be technically skilled.

   GAs are not suitable for the automatic search of large databases with a large number of candidate variables since GA software tends to be slow because the process of evaluation of the fitness function tends to be time-consuming. when the database and number of variables are large. In such cases decision trees may be more appropriate. Constructing GAs can be quite time-consuming and many runs may be required in the fitting process. GA solutions are difficult to explain as they do not provide statistical measures to enable the user to understand why the technique arrived at a particular solution.

The knowledge discovery process

Data mining can be conceived as the knowledge discovery process (KDP). Peter Peacock (2000) provides a model of KDP and the exposition below follows his model (see Exhibit 9.4). KDP is not new. It is the application of scientific discovery methods to large databases. The terms may be new, but the concepts are not. The elements in Exhibit 9.4 are described below. Although the exhibit does not have feedback loops, the KDP process is iterative in that there is a substantial flow of information back to prior steps in the process. Although KDP is generally discussed in the context of data mining, it is in fact common to all model building.

Data funnelling

Data mining techniques assume that the data quality is high. Data funnelling is a set of procedures that ensure that the data collected are suitable for analysis. They include the identification of internal operational data, appropriate external data, moving them to a data repository, evaluating data quality, and obtaining better data when necessary. Data quality is assessed by running simple queries, applying basic visualization techniques, and running automatic validation procedures. The objective here is not to make the data flawless – a near-impossible objective when we deal with large databases – but rather to make sure that there are no gross errors such as wrong type of data, outliers that are clearly wrong, and right data in the wrong column. Data

EXHIBIT 9.4

The knowledge discovery process

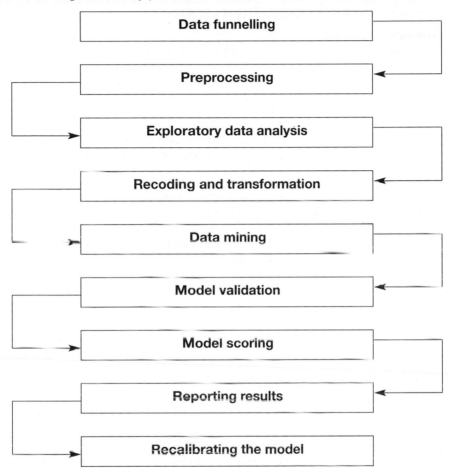

funnelling also includes choosing the subset of variables to be analysed from the larger set of all characteristics available in the data repository.

Preprocessing

The next step, data preprocessing, includes the following aspects:

1. *Reformatting*. Formatting data from different sources to a common format.
2. *Standardizing*. Standardizing data attributes to conform to a standard. For example, an organization may have standard specifications for an attribute. The data may have to be converted to conform to this standard. This is particularly true of text-based attributes.
3. *Removing records with sparse data*. Removing records with insufficient information for analysis purposes.
4. *Removing duplicate records.*

5. *'Householding'*. When the target unit for the analysis is a household rather than an individual, individuals must be assigned to households. This operation is generally performed by software that looks for sets of common variables such as last names, address components, and phone numbers.

Exploratory data analysis

Exploratory data analysis is used to identify the anomalies and outliers that remain in the data set after it has passed through the previous checks. It also provides the researcher with a 'feel' for the preprocessed data through ranges, means, measures of central tendency and dispersion, shape of the distributions, and correlations among variables. The analyst looks for largest and smallest values, central tendency, dispersion, the shapes of the distributions of individual variables, and the structure of the relationships among variables. This step often enables the analyst to generate preliminary hypotheses with regard to the nature of relationships among variables.

Recoding and transformation

In this phase, additional operations are performed on the data. This may include recoding the data to conform to the analyst's hypotheses, creating new variables by combining the existing variables, and the application of data transformation to non-linear data. Data may also be recoded into other values using simple decision rules. Recoding is used to convert continuous data to a nominal form for use with tools such as neural nets and decision trees. It can also be used to convert nominal text label data into numeric values.

Data mining

This phase includes techniques of machine learning from patterns in data using major discovery tools, association rules, decision trees, neural nets, and genetic algorithms.

Model validation

Once the model is built, it is important to assess its validity because models that might have worked on a training set may not work very well when applied to other data. A common approach to model validation is to draw two random samples from the preprocessed data: a 'calibration sample' and a 'holdout sample'. The calibration sample is used to build the model. This model is then tested against the holdout sample to validate the model. If the model performs very poorly on the validation sample then the analyst must modify the model and even rebuild it from scratch.

Model scoring

Model scoring refers to the application of the model developed by the analyst to the entire database. It is done through a set of classification rules developed on the basis of the calibration sample. For example, an equation such as $y = a + b_1 x_1 + b_2 x_2$ developed from a sample of the data is applied to the entire population of records. The scores derived by applying the formula to the entire database, the ys, are placed in a new column in the data base. These are generally known as *scores*, and the process as *scoring*. Scoring may also refer to the process of identifying cluster membership of individual observations when cluster analysis is carried out.

Reporting the results

Once the above processes are completed, the researcher interprets the results and presents them to the decision-maker along with supporting information.

Recalibrating the model

Because marketing is a dynamic process, behaviour patterns identified today may not work two years from now. All models – especially in applied disciplines such as marketing – deteriorate over time, and should be recalibrated regularly, preferably at definite intervals established beforehand. Recalibrating is the process of rebuilding the model with a recently constructed data set. The recalibrated model may differ from the original in terms of weights applied to the attributes in the models, include new attributes or even have a completely different formulation.

9.3 Application of data mining in market research

Data mining can be applied to marketing data in a wide variety of ways. Such applications continue to expand as the following reading shows. In his article, Michael S. Garver (2002) applies data mining techniques to customer satisfaction measurement.

Using Data Mining for Customer Satisfaction Research

Customer satisfaction research has become commonplace over the last 20 years, with businesses and academic researchers touting continuous improvement strategies driven by customer satisfaction data. Traditionally, researchers have used statistical techniques to analyze customer satisfaction data. Yet traditional techniques have limitations, especially in customer satisfaction research. In practice, many researchers ignore assumptions and limitations, which may produce biased and misleading results.

Data mining techniques have recently gained popularity with researchers, in part because they overcome many limitations of traditional statistics and can handle complex data sets. New data mining applications are appearing at an explosive rate, but they haven't focused on customer satisfaction research. Data mining techniques offer a powerful complement to statistical techniques and have useful research applications for customer satisfaction. They can also help answer typical customer satisfaction research questions.

What's the relative importance of customer satisfaction attributes?

When evaluating continuous improvement initiatives, researchers often study the importance customers place on different product and service attributes. Some directly ask respondents to rate the importance of different attributes, but most researchers use statistically inferred ratings of importance. Employing statistical methods, researchers regress ratings of customer satisfaction with different attributes (independent variables) against overall satisfaction (dependent variable). Typically, multiple regression accomplishes this task.

Multiple regression and related techniques make numerous assumptions that are often violated in practice. A typical problem in customer satisfaction research pertains to high levels of correlation between attributes (multicollinearity). This can dramatically affect the standardized beta coefficients, the statistical value used to determine relative importance of various attributes. If high levels of multicollinearity exist, then standardized beta coefficients probably will be biased, leading to inaccurate importance ratings. Furthermore, multiple regression assumes a normal distribution of ratings (i.e., that the scores will resemble a normal bell curve). This isn't the case with customer satisfaction. Past research has shown that data about customer satisfaction is often positively skewed. The majority of satisfaction scores fall at the upper end of the scale (8 to 10 on a 10-point scale). Finally, these statistical techniques also assume linear relationships between independent and dependent variables. This too is a mistake. Research has clearly demonstrated that relationships are often curvilinear, far from a straight line. In many industries, statistical assumptions don't hold and can result in biased and misleading results.

What's the impact of satisfaction on future financial performance?

Satisfaction research seeks to link measures of customer satisfaction to financial performance. To accomplish this goal, researchers use regression analysis or structural equation modeling, which can display sensitivity (ROI) analysis for different improvement initiatives. For example, these statistical techniques can show how satisfied customers would be with certain changes and how these changes would affect overall satisfaction and financial performance (Rust, Zahorik, and Keiningham, 1995). Accurate analysis of this kind is extremely valuable, yet invalid statistical assumptions affect the accuracy of sensitivity analysis. Businesses that make these errors think the results are precise when really they are not.

What level of attribute performance should be set as a target?

Setting performance targets for customer satisfaction attributes is extremely difficult for many practitioners. Often, managers arbitrarily set them. Brandt (1999) suggests taking an outside-in approach to performance targets. He proposes that attribute performance targets should be set where important business goals such as sales revenue, market share, and retention are realized. For example, if on-time delivery with a score of 8 leads to loyalty, then the performance target for on-time delivery should be 8. Brandt suggests a focus on appropriate levels of 'upstream performance' so that desired 'downstream performance' is obtained.

To develop performance targets, Myers (2001) suggests using performance breaks. This analysis graphs the relationship between the performance of a single attribute and a desired business outcome such as overall satisfaction, sales revenue, or retention. While the method is helpful, it lacks rigor because it can lead to a variety of interpretations. Better methods are needed.

How it works

Data mining techniques overcome many limitations of traditional analysis and statistical techniques. Neural networks and decision trees are two useful data mining techniques.

Neural network analysis

Since the mid-1980s, neural networks have been the focus of a great deal of research. With the advent of high-power computing and improved algorithms, neural network analysis is now commonplace. Originally developed in the 1940s, the term 'neural networks' refers to models designed to simulate the human brain. As an analytic tool, neural networks overcome the limitations of traditional statistics, which are typically visible in customer satisfaction research. Neural networks are mathematically driven. Bishop (1994) has shown how, in situations where non-normal data, multicollinearity, and non-linear relationships are present, neural networks will outperform multiple regression.

Multiple regression cannot cope with high levels of complexity. For example, it can only handle a limited number of variables, often no more than five. In contrast, neural networks and other data mining techniques can handle up to 200 variables and are thus far more realistic in representing the complexity which is typical of the business environment.

Although researchers regularly draw elaborate comparisons of neural networks to the human brain, neural networks provides similar results to multiple regression. Although these techniques analyze the data using different methods, conceptually, the results are similar. Multiple regression uses independent variables to predict a dependent variable. Neural networks call variables neurons and include one or more neurons in an input layer (independent variables) to predict one or more neurons in an output layer (dependent variables). Both techniques are designed to make predictions and examine the impact of variables on those predictions.

The key to neural network analysis as a model is the way it handles complexity and uncertainty. The hidden neurons make the connection between input and output layers indirect. This is intended to account for and model complexity and uncertainty in the business world. This use of hidden neurons reflects an actual level of uncertainty and complexity found in business operations.

The simplest form of neural networks consists of three layers. The input layer includes one or more neurons that are independent (predictor) variables. The output layer includes one or more neurons that are dependent (outcome) variables. A hidden layer connects these two layers and their neurons. Similar to multiple regression, the input neurons predict the output neuron.

There are two important considerations that affect neural network analysis: (1) over-training and (2) random starting positions. Most software packages include features to prevent over-training, which is critical to preventing memorization of the data, producing results that will not generalize to the population. Random starting positions are necessary because starting positions will influence the results. Similar results generated by a number of random starting positions provide confidence in the results.

Neural network applications

Neural network analysis is well suited for research into customer satisfaction. It can evaluate the relative importance of customer satisfaction attributes and predict the effect of customer satisfaction on future financial performance (sensitivity analysis). Data from pizza customers illustrate the use of data mining techniques and how they compare to traditional approaches. Because financial data wasn't available, only relative importance will be analyzed. Yet, with the inclusion of financial data, a similar model could analyze the effect of different issues on financial performance and evaluate the effect of proposed changes.

Data about customer satisfaction were collected from customers of a national pizza chain. The questionnaire inquired about customer perceptions of the pizza they ate and about their overall satisfaction with the pizza and the service of the chain. The customers were asked about: (1) taste, (2) amount of pizza, (3) speed of delivery, (4) reliability of delivery, (5) price, (6) delivered temperature, and (7) friendliness of employees. A 10-point, semantic differential scale (1 = poor, 10 = excellent) was used. Both multiple regression and neural network analysis were conducted to rate the relative importance of the seven attributes. In multiple regression analysis, attributes were entered as independent variables and overall satisfaction as the dependent variable. SPSS 10.0 was used to conduct the analysis.

In the neural network analysis, these same pizza attributes were entered as neurons in the input layer. Overall customer satisfaction was entered as the neuron in the output layer. Given the number of input neurons, three neurons were included in the hidden layer. Clementine 6.0, the SPSS data mining software package, was used to analyze the data.

The importance rating of different attributes was determined using sensitivity analysis in Clementine. In the past, a major criticism of neural networks has been the inability to explain their reasoning – resembling a black box. Sensitivity analysis overcomes this problem and provides scores, which measure the importance of each neuron in the input layer. The scores for pizza attributes are derived from their ability to predict overall satisfaction and indicate the relative importance of each attribute, ranging between 0 and 1, with higher values representing more importance. Because this data set does not meet the conditions necessary for effective statistical analysis, neural network analysis provides more accurate ratings of attribute importance. The results from multiple regression analysis and neural network analysis are provided in Exhibit 1.

EXHIBIT 1

Relative importance analysis

Pizza attributes	Multiple regression*	Neural networks**
Taste	.52 (1)***	.55 (1)
Amount	.10 (4)	.42 (2)
Speed of delivery	.01 (6)	.14 (5)
Reliable delivery	.13 (3)	.31 (4)
Price	.28 (2)	.41 (3)
Delivered temperature	.06 (5)	.02 (7)
Friendliness	−.06 (7)	.07 (6)

* Values in this column represent standardized beta coefficients.
** Values in this column represent impact scores from sensitivity analysis.
***Numbers in parentheses represent the attributes' rank order of importance.

Both techniques identified taste as the most important pizza attribute, but neural networks identified amount as the second most important attribute, while multiple regression places this variable as a distant fourth. Regression analysis displays a negative coefficient for friendliness, and two variables (taste and price) have much higher coefficients than the remaining pizza attributes. This is likely due to high levels of multicollinearity. These different techniques suggest quite different views of what is important.

Decision tree analysis

Decision tree analysis is another popular data mining technique, also driven by a mathematical algorithm. As with neural networks, non-normal distributions, multicollinearity, and non-linear relationships do not hinder the performance of decision tree analysis. In regard to complexity, decision tree analysis can handle up to 200 hundred predictor variables.

Decision tree analysis, also called rule induction, uses different algorithms including (1) CHAID, (2) C5.0, and (3) C&RT. Each technique is designed for slightly different purposes. For example, C5.0 can only predict categorical dependent variables, whereas C&RT can be used to predict interval or continuous variables.

Essentially, decision tree analysis searches through the data to identify which predictor variable is most important to correctly predicting the dependent variable. Decision tree analysis partitions the data based on an initial split of the first variable. Thus, the most important predictor is selected first. For that variable, the level of that variable that corresponds to predicting the dependent variable is also identified. Then, the next most important variable for predicting the dependent variable is selected, along with the corresponding level of performance. Decision tree analysis continues until all relevant variables are selected. If a variable is not selected, then it is not critically important to the prediction.

Decision tree analysis may also yield insight into different market segments. For example, the loyalty of different market segments is affected by different attributes in different ways. Decision tree analysis can identify various market segments in the data, display what is most important for that segment, and define the level of performance that would create loyalty.

Decision tree applications

Decision tree analysis can be used to rank the importance of various attributes and to rigorously identify performance targets for customer satisfaction attributes. According to Brandt's outside-in approach, attribute performance levels should be set so that desired business outcomes are achieved. Assuming that customer loyalty is the desired business outcome, decision tree analysis will rank the importance of customer satisfaction attributes according to how they predict loyalty and will indicate the level of performance required to create loyalty.

Decision tree analysis is illustrated using data from the pizza chain about customer satisfaction. In this analysis, loyalty is the desired business outcome (dependent variable) and the pizza and service attributes are the predictor (independent) variables. The results are displayed in Exhibit 2.

EXHIBIT 2

Identifying performance targets

Pizza attributes	Loyalty segment A Accuracy = .971	Loyalty segment B Accuracy = .915	Non-loyal segment Accuracy = .925
Taste	8 (1)*	9 (1)	7 or below
Amount		9 (2)	8 or below
Speed of delivery			
Reliable delivery	8 (3)		
Price	9 (2)		
Delivered temperature			
Friendliness			

* The first number is the minimum performance requirement to become loyal, whereas the number in parentheses refers to its rank-order importance to that segment.

The results of this analysis tell an interesting story. Two segments show the levels of performance that creates loyalty. One segment tells us how we have destroyed loyalty in the past. The loyalty of segment A perceives attribute importance in the following descending order: (1) taste, (2) price, and (3) reliable delivery. This segment requires the following minimum performance levels to become loyal: taste (8), price (9), and reliable delivery (8). Segment A wants a great-tasting pizza (8 or better) that is very affordable (9 or better), and one that will be delivered reliably (8 or better). In contrast, segment B wants a great-tasting pizza (9 or better) that's big in size (9 or better). For these two segments, the predictive models are very accurate with levels of .971 and .915. If pizza customers perceive taste to be 7 or below, and the amount to be 8 or below, then there's a high likelihood (.925) these customers will not be loyal.

The results of decision tree analysis are similar to both regression and neural network analysis and help confirm the results. Additionally, decision tree analysis goes beyond both of these techniques by providing insight into different segments and their perceptions of both attribute importance and performance. Clearly, researchers can gain the most understanding by analyzing customer satisfaction data with both statistical and data mining techniques.

Ensuring success

Many researchers perceive data mining techniques as tools to 'fish' for results, with no regard for theory or validation. Rigorous data mining is far from a weekend fishing trip. It's critical that researchers follow a rigorous data mining methodology. The CRISP-DM methodology was created to do just that. (An in-depth discussion of the CRISP-DM methodology can be reviewed at http://www.crisp-dm.org.) A critical element of the CRISP-DM methodology concerns the validation of models. Data mining techniques are not based on statistical theory, thus validating and testing models is imperative. Once a model is developed on practice data, the model is validated on a holdout sample. Using business knowledge and theory as a guide, models are created and tested in one data set, then validated on another sample. These tests are the only way to ensure accurate, valid models that generalize to the population. Typically, the training set represents about 60% of the sample, while the test and validation samples represent 30% and 10%, respectively. Researchers should expect a small drop in performance on both the test and validation samples, yet the models should perform relatively well. This practice ensures that the model produced is not valid only in the training data, but will generalize to the population.

Researchers have long known that research triangulation offers the most valid and reliable perspective of customers. Different methods and techniques should be used as complements to draw conclusions about customers. Neural networks and decision tree analysis do not replace but complement statistical techniques. Researchers should look for convergence among both statistical and data mining techniques. If neural networks, decision trees, and regression analysis all derive similar results, then researchers can have increased confidence in the results. However, in the face of multicollinearity, non-normal distributions, and/or non-linear relationships, neural networks and decision trees are clearly the preferred analytic techniques.

Numerous data mining software programs are now available. Kdnuggets (www.Kdnuggets.com) is an online resource for data miners that has identified the most popular data mining software packages. The most popular vendors and their data mining software programs include the following:

- SPSS, www.spss.com/clementine
- SAS, www.sas.com/products/miner/index.html
- IBM, www-3.ibm.com/software/data/bi
- Angoss, www.angos.com
- Megaputer, www.megaputer.com

Each of these programs can run on client–server platforms and includes a wide range of analysis tools and techniques, including neural networks and decision trees. Processing time is dependent on the number of variables and the size of the data set, yet client–server platforms allow for complex processing of large data sets in a short amount of time.

Additional reading

Bishop, Chris M. (1994) Neural networks and their applications. *American Institute of Physics*, **65**, 1803–30.

Brandt, D. Randall (1999) *Customer Satisfaction Management Frontiers II*. Quality University Press, Fairfax, VA.

Myers, James (2001) *Measuring Customer Satisfaction: Hot Buttons and Other Measurement Issues*. American Marketing Association, Chicago.

Rust, Roland T., Zahorik, Anthony J. and Keiningham, Timothy L. (1995) Return on quality (ROQ): Making service quality financially accountable. *Journal of Marketing* **59**, 58–70.

Bibliography

Further reading

The following book is comprehensive and covers statistical modelling and analysis for database marketing. It contains a wide variety of applications using techniques discussed in the present volume.

Ratner, Bruce (2003) *Statistical Modeling and Analysis for Database Marketing*. Chapman & Hall/CRC, Boca Raton, FL.

References

Garver, Michael S. (2002) Using data mining for customer satisfaction research. *Marketing Research*, **14**(1), 8–12.

Peacock, Peter (2000) Database and data mining techniques. In Chuck Chakrapani (ed.), *Marketing Research: State-of-the-Art Perspectives*. American Marketing Association, Chicago, and Professional Marketing Research Society, Toronto.

Appendix
Technical Descriptions of Techniques

A.1 Basic statistical computations

Summarizing the data

For n individuals on p variables, the data matrix $\mathbf{x}$ is of the form

$$\mathbf{X} = \begin{pmatrix} x_{11} & x_{12} & \cdots & x_{1p} \\ x_{21} & x_{22} & \cdots & x_{2p} \\ \cdots & \cdots & \cdots & \\ x_{n1} & x_{n2} & \cdots & x_{np} \end{pmatrix}$$

with each row of the matrix representing an individual and each column a variable. The mean and variance for this matrix are given by:

Measure	Population	Estimate
Mean vector	$\boldsymbol{\mu}' = (\mu_1, \mu_2, \ldots, \mu_p)$	$\bar{\mathbf{X}}' = (\bar{x}_1, \bar{x}_2, \ldots, x_p)$
for variable i	$\mu_i = \mathrm{E}\,(x_i)$	$\bar{x}_i$ = sample mean
Variance vector	$\boldsymbol{\sigma}' = (\sigma_1^2, \sigma_2^2, \ldots, \sigma_p^2)$	$\mathbf{s}' = (s_1^2, s_2^2, \ldots, s_p^2)$
for variable i	$\sigma_i^2 = \mathrm{E}\,(x_i - \mu_i)^2$	s_i^2 = sample variance

Computing sums of squares

The raw sum of squares is defined as

$$\mathbf{B} = \mathbf{X}'\mathbf{X}.$$

The mean corrected scores are given by

$$\mathbf{X}_\mathrm{d} = \mathbf{X} - 1\,\bar{\mathbf{X}}'$$

Where $1\bar{\mathbf{X}}'$ denotes the matrix product of a unit column vector and the row vector $\bar{\mathbf{X}}'$.

The mean corrected sum of squares and cross-products matrix $\mathbf{S}$ is given by

$$\mathbf{S} = \mathbf{X}'_d \mathbf{X}_d.$$

Analysing relationships among variables

One of the basic purposes of multivariate analysis is to understand the relationship among different variables (and/or objects). The relationship between any two variables is measured through covariance and correlation.

The covariance between two variables x_i and x_j is defined as:

$$\text{Cov}(x_i, x_j) = \text{E}(x_i - \mu_i)(x_j - \mu_j).$$

The variance values of p variables can be arranged into a $p \times p$ square matrix $\mathbf{\Sigma}$, given by

$$\mathbf{\Sigma} = \begin{pmatrix} \sigma_{11} & \sigma_{12} & \cdots & \sigma_{1p} \\ \sigma_{21} & \sigma_{22} & \cdots & \sigma_{2p} \\ \cdots & \cdots & \cdots & \cdots \\ \\ \sigma_{p1} & \sigma_{p2} & \cdots & \sigma_{pp} \end{pmatrix}$$

which is a square symmetric matrix with $\sigma_{ij} = \sigma_{ji}$. The matrix is estimated by

$$\mathbf{S} = \mathbf{\Sigma}(\mathbf{x}_i - \bar{\mathbf{x}}'_i)(\mathbf{x}_j - \bar{\mathbf{x}}'_j)'/(n-1) = \mathbf{X}'_d \mathbf{X}_d/(n-1), \qquad i = 1, \ldots, n,$$

where $\mathbf{x}'$ is the vector of observations $x_{i1}, \ldots, x_{ip}$ for the ith individual.

The correlation coefficient ρ_{ij} is obtained by standardizing the covariance measure:

$$\rho_{ij} = \sigma_{ij}/(\sigma_{ii} \sigma_{jj})^{\frac{1}{2}}.$$

Correlation coefficient values range from -1.0 to $+1.0$. Positive correlation between variables x_i and x_j indicate that higher values of x_i are associated with higher values of x_j. Negative correlation between variables x_i and x_j indicate that higher values of x_i are associated with lower values of x_j.

The correlation coefficients for a set of variables can be arranged in a matrix. The diagonals of the matrix contain the correlation of each variable with itself, which is unity. The matrix may be written as

$$\mathbf{R} = \mathbf{D}^{-\frac{1}{2}} \mathbf{\Sigma} \mathbf{D}^{-\frac{1}{2}}$$

where $\mathbf{D}^{-\frac{1}{2}} = \text{diag}(1/\sqrt{\sigma_{11}}, \ldots, 1/\sqrt{\sigma_{pp}})$.

Linearly combining data

Multivariate analysis often involves creating new variables that are linear (weighted) combinations of existing variables. Given p variables $x_1, \ldots, x_p$ and a set of p scalars $a_1, \ldots, a_p$, we can create a linear combination $y = a_1 x_1 + a_2 x_2 + \ldots, a_p x_p$ or

$$y = \mathbf{a}'\mathbf{x} = \mathbf{x}'\mathbf{a},$$

where $\mathbf{a}' = (a_1, a_2, \ldots, a_p)$, with a mean of $\mathbf{a}'\mathbf{\mu}$ and variance of $\mathbf{a}'\mathbf{\Sigma}\mathbf{a}$.

A.2 Principal components analysis

The first principal component of the observations, f_1, is the linear combination of p variables such that sample variance is greatest among all such linear combinations,

$$f_1 = a_{11}x_1 + a_{12}x_2 + \ldots + a_{1p}x_p,$$

subject to

$$\sum_{i=1}^{p} a_{1i}^2 = 1.$$

The coefficients $(a_{11}, a_{12}, \ldots, a_{1p})$ can be written as the vector $\mathbf{a}'$.

The second principal component f_2 is the linear combination of p variables which has the greatest variance,

$$f_2 = a_{21}x_1 + a_{22}x_2 + \ldots + a_{2p}x_p = \mathbf{a}_2'\mathbf{x},$$

subject to the following constraints:

$$\mathbf{a}_j'\mathbf{a}_j = 1,$$
$$\mathbf{a}_j'\mathbf{a}_i = 0.$$

With p variables, there can be p components.

To find the coefficients of the vector $(a_{11}, a_{12}, \ldots, a_{1p})$, we need to choose the elements of the vector $\mathbf{a}_1$ so as to maximize the variance of f_1 subject to the constraint $\mathbf{a}_1'\mathbf{a}_1 = 1$. The variance of f_1 is given by

$$\mathrm{Var}(f_1) = \mathrm{Var}(\mathbf{a}_1'\mathbf{x}) - \mathbf{a}_1'\mathbf{S}\mathbf{a}_1.$$

The use of the method of Lagrange multipliers leads to the solution that $\mathbf{a}_1$ is the eigenvector of $\mathbf{S}$ corresponding to the largest eigenvalue. This procedure is repeated to derive the remaining components, with $\mathbf{a}_j$ being the eigenvector of $\mathbf{S}$ associated with the jth largest eigenvalue.

If the eigenvalues of $\mathbf{S}$ are $\lambda_1, \lambda_2, \ldots, \lambda_p$, then the variance of the ith principal component is λ_i (since $\mathbf{a}_i'\mathbf{a}_i = 1$). An important property of eigenvalues is that the sum of the variances of the principal components is equal to the sum of the variances of the original variables. The components account for all the variance in the original data.

$$\sum_{i=1}^{p} \lambda_i = \mathrm{tr}\,(\mathbf{S}).$$

The proportion of variance accounted for by the jth principal component is given by

$$P_j = \frac{\lambda_j}{\mathrm{tr}\,(\mathbf{S})}$$

The covariance of the original variables with the jth principal component is given by:

$$\mathrm{Cov}\,(\mathbf{x}, f_j) = \mathrm{Cov}(\mathbf{x}, \mathbf{x}'\mathbf{a}_j)$$
$$= \mathrm{E}\,(\mathbf{x}\mathbf{x}')\mathbf{a}_j$$
$$= \mathbf{S}\mathbf{a}_j$$
$$= \lambda_j\,\mathbf{a}_j.$$

The covariance and correlation of variable i with component j are given by

$$\text{Cov}(x_i, y_j) = \lambda_j a_{ji},$$

$$r_{xi,yi} = \frac{\text{Cov}(x_i, y_j)}{\sigma_{xi}\sigma_{yi}}$$

$$= \frac{\lambda_j a_{ji}}{s_{ii}^{\frac{1}{2}}\sqrt{\lambda_j}}$$

$$= \frac{\sqrt{\lambda_j}a_{ji}}{s_{ii}^{\frac{1}{2}}}$$

Consequently, if the components are extracted from the correlation matrix rather than the covariance matrix, then

$$r_{xi,yi} = \sqrt{\lambda_j}a_{ji}.$$

A.3   Common factor analysis

In common factor analysis, as in principal components analysis, we start with a set of observed variables, $\mathbf{x}' = (x_1, x_2, \ldots, x_p)$, but assume that they can be described as a manifestation of a smaller number of variables or factors $f_1, f_2, \ldots, f_k$ where $k < p$. Thus each observed variable is described in terms of a linear (regression-like) model:

$$x_1 = \lambda_{11}f_1 + \lambda_{12}f_2 + \ldots + \lambda_{1k}f_k + e_1,$$

$$x_2 = \lambda_{21}f_1 + \lambda_{22}f_2 + \ldots + \lambda_{2k}f_k + e_2,$$

$$\ldots$$

$$x_p = \lambda_{p1}f_1 + \lambda_{p2}f_2 + \ldots + \lambda_{pk}f_k + e_k,$$

Or, using the matrix notation,

$$\mathbf{x} = \mathbf{\Lambda f} + \mathbf{e},$$

$$\mathbf{\Lambda} = \begin{pmatrix} \lambda_{11} & \cdots & \lambda_{1k} \\ \cdots & \cdots & \cdots \\ \lambda_{p1} & \cdots & \lambda_{pk} \end{pmatrix}, \quad \mathbf{f} = \begin{pmatrix} f_1 \\ \cdots \\ f_k \end{pmatrix}, \quad \mathbf{e} = \begin{pmatrix} e_1 \\ \cdots \\ e_k \end{pmatrix}.$$

Here the 'error' or the 'residual' terms $e_1, \ldots, e_p$ are assumed to be uncorrelated with each other and with the factors $f_1, \ldots, f_k$. The regression coefficients in $\mathbf{\Lambda}$ are the *factor loadings*.

If we express the factors as standardized latent variables with a mean of zero and a standard deviation of 1 and assume that they are uncorrelated with one another, the variance of variable x_i given by:

$$\sigma_i^2 = \sum_{j=1}^{k} \lambda_{ij}^2 + \epsilon_i^2,$$

where e_i^2 is the variance of e_i.

The variance associated with each variable then comes from two sources: variance that is unique or specific only to that variable (specific variance or ε_i); and variance shared with other variables through common factors (known as the communality or h^2), which is given by

$$h_i^2 = \sum_{j=1}^{k} \lambda_{ij}^2.$$

The covariance of observed variables x_i and x_j is given by:

$$\sigma_{ij}^2 = \sum_{l=1}^{k} \lambda_{il}\lambda_{jl}.$$

As in principal components analysis, when the factors are uncorrelated or *orthogonal*, the factor loadings are simply the correlations between factors and observed variables.

The covariance matrix, Σ, of the observed variables is given by

$$\Sigma = \Lambda\Lambda' + \psi$$

where $\psi = \text{diag}(\varepsilon_i)$.

In practice, Σ is estimated by the sample covariance matrix S (or the correlation matrix R), and we will need to obtain estimates of Λ and ε so that the observed covariance matrix takes the form required by the model.

A.4   Correspondence analysis

Our data table is a matrix K with n rows and p columns. A typical element of this matrix is written as k_{ij}. The total of the row i elements is written as k_i.

$$k_i = \sum_{i=1}^{p} k_{ij},$$

and the total of the column j elements is written as $k_{.j}$

$$k_{.j} = \sum_{j=1}^{n} k_{ij}.$$

The total for the table is written

$$k = \sum_{i,j} k_{ij}.$$

The n row profiles are considered as n points (or vectors) in a p-dimensional space $\mathbb{R}^p$. The jth component of the ith vector is

$$\frac{k_{ij}}{k_j} \text{ for } j = 1, 2, ..., p.$$

The p column profiles are considered as points (or vectors) in an n dimensional space $\mathbb{R}^n$. The ith component of the jth vector is

$$\frac{k_{ij}}{k_{.j}} \text{ for } i = 1,2, \ldots, n.$$

In terms of relative frequencies, if each number k_{ij} of the data matrix is divided by the total k for the table, then we get a table **F** of relative frequencies f_{ij}:

$$f_{ij} = \frac{k_{ij}}{k.}$$

The mass of row i is written

$$f_{i.} = \frac{k_{i.}}{k,}$$

and the mass of column j is written

$$f_{.j} = \frac{k_{.j}}{k,}$$

and we have

$$\frac{f_{ij}}{f_{.j}} = \frac{k_{ij}}{k_{.j}} \text{ and } \frac{f_{ij}}{f_{.j}} = \frac{k_{ij}}{k_{i.}} \text{ for all } i \text{ and } j.$$

Distance

The (squared) distance between two row points is defined in correspondence analysis as follows:

$$d^2(i,i') = \sum_{j=1}^{p} \frac{1}{f_{.j}} \left(\frac{f_{ij}}{f_{i.}} - \frac{f_{i'j}}{f_{i'.}} \right)^2.$$

In a symmetrical way the distance between two column profiles is defined as:

$$d^2(j,j') = \sum_{i=1}^{n} \frac{1}{f_{i.}} \left(\frac{f_{ij}}{f_{.j}} - \frac{f_{ij'}}{f_{.j'}} \right)^2.$$

This distance differs from the usual Euclidean distance only in the fact that each squared term is weighted by the inverse of the frequency corresponding to each term. We note two properties of our distance measure:

1. If two rows having identical profiles are grouped together into a single row, the distances between the columns remain unchanged.

2. If two columns having identical profiles are grouped together into a single column, the distances between the rows remain unchanged.

These properties ensure a certain invariance of the results vis-à-vis the nomenclature adopted.

Consider rows and columns as points in multidimensional spaces. If $\mathbf{D}_p$ is the diagonal $p \times p$ matrix whose jth diagonal element is $f_{.j}$, and if $\mathbf{D}_n$ is the diagonal $n \times n$ matrix whose ith diagonal element is $f_{i.}$, then in $\mathbb{R}^p$ there are n points whose coordinates are given by the n rows of the matrix $\mathbf{D}_n^{-1}\,\mathbf{F}$ with masses $f_{i.}$, and in $\mathbb{R}^n$ there are p points whose coordinates are given by the p columns of $\mathbf{F}\,\mathbf{D}_p^{-1}$ (or the p rows of $\mathbf{D}_p^{-1}\mathbf{F}'$), with masses $f_{.j}$.

Analysis in $\mathbb{R}^p$

In this space, the n points are the n rows of $\mathbf{D}_n^{-1}\mathbf{F}$, with masses $f_{i.}$. The directions of the principal axes of inertia and the inertias associated with them are given by the eigenvectors $\mathbf{u}$ and the eigenvalues λ associated with them, which are the solutions of the equation

$$\mathbf{Su} = \lambda\mathbf{u}$$

where $\mathbf{S} = \mathbf{F}'\,\mathbf{D}_n^{-1}\mathbf{F}\,\mathbf{D}_p^{-1}$.

The general term of the matrix $\mathbf{S}$ is

$$s_{jj'} = \sum_{i=1}^{n} \frac{f_{ij}f_{ij'}}{f_{i.}f_{.j'}}.$$

The first principal axis $\mathbf{u}_1$ is the one associated with the highest value of λ_1. The second principal axis $\mathbf{u}_2$ is the one associated with the next highest value of λ_2 and so on. In correspondence analysis the maximum number of principal axes is the smaller of the numbers $n-1$ and $p-1$.

The principal coordinates of the n points on a principal axis $\mathbf{u}_\alpha$ are given by

$$\mathbf{D}_n^{-1}\mathbf{F}\,\mathbf{D}_p^{-1}\,\mathbf{u}_\alpha.$$

Analysis in $\mathbb{R}^n$

It is never necessary to perform this analysis because in correspondence analysis the same number of principal axes are obtained in both analyses and they each account for exactly the same amounts of inertia in each analysis. (In this space the principal axes and the inertias accounted for by them will be the eigenvectors $\mathbf{v}$ of

$$\mathbf{F}\,\mathbf{D}_p^{-1}\mathbf{F}'\,\mathbf{D}_n^{-1}$$

and the eigenvalues associated with them.) Also because of a duality relationship that exists between the principal coordinates of the row points and the principal coordinates of the column points, knowing the coordinates of the row points, we can compute those of the column points, and vice versa, by what is known as the *transition formula*.

Transition formula

If the coordinate of the row point i on the principal axis α is written $\mathbf{F}_\alpha (i)$ and the coordinate of the column point j on the principal axis α is written $\mathbf{G}_\alpha (j)$, then

$$\mathbf{F}_\alpha (i) = \sqrt{\frac{1}{\lambda_\alpha}} \sum_{j=1}^{p} \frac{f_{ij}}{f_{i.}} \mathbf{G}_\alpha (j)$$

and

$$\mathbf{G}_\alpha (j) = \sqrt{\frac{1}{\lambda_\alpha}} \sum_{i=1}^{n} \frac{f_{ij}}{f_{.j}} \mathbf{F}_\alpha (j)$$

A perceptual map is obtained by superimposing the plane of the first two axes of one analysis on the plane of the first two axes of the second analysis. Similarly, any two axes of one analysis are superimposed on the corresponding axes of the other analysis to obtain additional information from the subsequent axes.

A.5   Regression analysis

Given p independent variables (xs) and a dependent variable (y), the regression equation is of the form

$$\hat{y} = b_0 + b_1 x_1 + b_2 x_2 + \ldots + b_p x_p,$$

which is the estimate of the true but unknown population equation

$$\mu = \beta_0 + \beta_1 x_1 + \beta_2 x_2 + \ldots + \beta_p x_p.$$

We model each score y_i as a linear function of bs. Thus

$$y_i = b_0 + b_1 x_{i1} + b_2 x_{i2} + \ldots + b_p x_{ip} + e_i.$$

Conversely

$$e_i = y_i - \hat{y}_i$$

In matrix notation,

$$\mathbf{y} = \mathbf{Xb} + \mathbf{e}.$$

The least-squares estimates of the bs are given by

$$\hat{\mathbf{b}} = (\mathbf{X'X})^{-1} \mathbf{X'y}.$$

Breakdown of sum of squares

The total sum of squares about the mean (SS) can be broken down into two component parts: the sum of squares due to regression (SS_{REG}) and sum of squares about regression (SS_{RES}).

$$\sum (y_i - \bar{y})^2 = \sum (y_i - y')^2 + \sum (y_i' - \bar{y}')^2$$

sum of squares about mean	sum of squares about regression	sum of squares due to regression	
SS_T	$= SS_{RES}$	$+ SS_{REG}$	
$n-1$	$= n-p-1$	$+ p$	(degrees of freedom)

This leads to the following analysis variance table:

Source	SS	df	MS	F
Regression	SS_{REG}	p	SS_{REG}/p	
Residual	SS_{RES}	$n-p-1$	$SS_{RES}/(n-p-1)$	MS_{REG}/MS_{RES}

The coefficient determination, or R^2, which measures the proportion of variance accounted for by the set of predictors given by

$$R^2 = SS_{REG}/SS_T.$$

This can be rewritten in terms of the F ratio:

$$F = \frac{R^2/p}{(1 - R^2)/(n-p-1)}$$

with p and $n-p-1$ degrees of freedom.

A.6 Discriminant analysis

Two-group discriminant analysis

Given p independent variables (xs), the function that discriminates between two groups is given by

$$Z = a_1x_1 + a_2x_2 + \ldots + a_px_p$$

The coefficients $a_1, a_2, \ldots, a_p$ are chosen such that the ratio of the between-group variance of z to its within-group variance is maximized:

$$V = \mathbf{a'Ba}/\mathbf{a'Sa},$$

where $\mathbf{B}$ is the covariance matrix of the group means and $\mathbf{S}$ is the pooled within-groups covariance matrix. The vector $\mathbf{a}' = (a_1, a_2, ..., a_p)$ which maximizes V is obtained by solving the equation

$$(\mathbf{B} - \lambda\mathbf{S})\mathbf{a} = 0.$$

For two groups, the single solution can be shown to be

$$\mathbf{a} = \mathbf{S}^{-1}(\bar{x}_1 - \bar{x}_2).$$

Individuals are assigned to groups based on the cutoff value (C) which is defined as

$$C = \begin{cases} (\bar{Z}_1 + \bar{Z}_2)/2 & \text{for equal sample sizes,} \\ (n_1\bar{Z}_1 + n_2\bar{Z}_2)/(n_1 + n_2) & \text{for unequal sample sizes.} \end{cases}$$

More than two groups

When there are more than two groups a series of classification coefficients is derived for each pair of groups. For instance, if we have three groups, we have three pairs: 12, 13 and 23. The corresponding functions are

$$h_{12}(\mathbf{x}) = (\bar{\mathbf{x}}_1 - \bar{\mathbf{x}}_2)'\,\mathbf{S}^{-1}[\mathbf{x} - (\bar{\mathbf{x}}_1 + \bar{\mathbf{x}}_2)/2]$$

$$h_{13}(\mathbf{x}) = (\bar{\mathbf{x}}_1 - \bar{\mathbf{x}}_3)'\,\mathbf{S}^{-1}[\mathbf{x} - (\bar{\mathbf{x}}_1 + \bar{\mathbf{x}}_3)/2]$$

$$h_{23}(\mathbf{x}) = (\bar{\mathbf{x}}_2 - \bar{\mathbf{x}}_3)'\,\mathbf{S}^{-1}[\mathbf{x} - (\bar{\mathbf{x}}_2 + \bar{\mathbf{x}}_3)/2]$$

where $\mathbf{S}$ is the pooled within-groups covariance matrix computed over all three groups. Individuals are assigned to groups based on the following classification rules:

If $h_{12}(\mathbf{x}) > 0$ and $h_{13}(\mathbf{x}) > 0$, assign to G_1.

If $h_{12}(\mathbf{x}) < 0$ and $h_{23}(\mathbf{x}) > 0$, assign to G_2.

If $h_{13}(\mathbf{x}) < 0$ and $h_{23}(\mathbf{x}) < 0$, assign to G_3.

A.7 Conjoint analysis

Traditional conjoint analysis is based on Thurstone's random utility theory, which suggests that individuals choose an alternative that maximizes their utility (i.e., the one that they like best, subject to factors that constrain their choices).

The ranking or rating for a given alternative R_{ij} is a linear combination of explanatory variables whose rescaled coefficients are their utilities (called partworths, since when combined, they are the 'worth' or the total utility of the chosen alternative):

$$R_{ij} = \alpha\mathbf{x}_{ij}' + \varepsilon_{ij}$$

where the $\boldsymbol{\alpha}$ is the $K \times 1$ vector of utility coefficients $(a_1, a_2, ..., a_k)$ attached to K explanatory variables, $\mathbf{x}'$.

The partworth function is computed for each individual separately. Using dummy variable regression, the number of independent parameters to be estimated will be equal to:

$$\left(\sum_{k=1}^{K} (M_k - 1) \right) - 1$$

where K is the number of attributes and M_k the number of levels of attribute k.

The importance I of the kth attribute is given by

$$I_k = \max (x_k) - \min (x_k).$$

To assess the importance of a given attribute in relation to others, we can normalize it:

$$W_k = \frac{I_k}{\sum_{k=1}^{K} I_k}$$

so that

$$\sum_{k=1}^{K} w_k = 1.$$

Author Index

Subject Index